Chain Mail: Newsletter Correspondence on Bitcoin and Blockchain Tech

Published by Nova Luna LLC
Portsmouth, NH

Copy edited and proofread by Rachael Churchill

Library of Congress Control Number: 2023940341

Paperback: 979-8-9892298-0-2
Hardback: 979-8-9892298-1-9
eBook: 979-8-9892298-2-6

FIRST EDITION 2023

Printed in the USA

Chain Mail

Newsletter Correspondence
on Bitcoin and Blockchain Tech

Stevie Sats and
Maria Ortiz Spillane

For our families.

"People have been defending their own privacy for centuries with whispers, darkness, envelopes, closed doors, secret handshakes, and couriers. The technologies of the past did not allow for strong privacy, but electronic technologies do.

We the Cypherpunks are dedicated to building anonymous systems. We are defending our privacy with cryptography, with anonymous mail forwarding systems, with digital signatures, and with electronic money."

—Eric Hughes,
"A Cypherpunk's Manifesto" (1993)

Contents

Part II — Altcoins: You Win Some, You Lose Some

Part III — Web 3.0: Decentralizing the World Wide Web

Part IV — NFTs and the Metaverse: Brave New You

Contents Continued

Key Words in Crypto

Before you get started, here's a list of words that you might want to refer to while reading. They pop up throughout the book, and though we'll try our best to define them, it could help to have them all in one place. If not, tear this page out (carefully) and give it to a friend who needs a cheat sheet.

Altcoin: Any crypto that's not Bitcoin.

Blockchain: A technology that allows data to be stored in a series of blocks that are linked one after the other in a chain, each containing a hash of the data from before. This tech underpins forms of cryptocurrency like Bitcoin and Ethereum.

Consensus Mechanism (or Consensus Algorithm): The way in which the computers in a network must reach agreement in order to work together securely.

Bitcoin: The first viable digital currency; a network, protocol and cryptocurrency all in one.

Cypherpunks: People who advocate privacy tech as a means for social preservation and progress, usually through the use of public and private key cryptography.

dApps (Decentralized Applications): Software programs that run on a decentralized network of computers.

Digital Wallet: A method of storing a person's crypto or digital assets. A hot wallet stores the keys needed to access a user's crypto online (enabled by the internet), whereas a cold wallet stores keys offline (often on a USB drive or other physical instrument).

Double Spend Problem: Spending or transacting the same digital money more than once; fundamental issue that Satoshi solved in creating Bitcoin.

DYOR: An acronym that stands for "Do Your Own Research."

Extended Reality (XR): A combination of tech elements within virtual reality (first-person, simulated environment) and augmented reality (first-person, hybrid real or virtual space)

Fintech: A portmanteau combining the words "financial" + "technology."

Golden Cross: A trading signal that may indicate an upward turn in a market; analysts often cite this as a potential sign of a bull market ahead.

Immutable: Unchanging over time or unable to be changed.

Lightning Network: A second-layer solution that aims to make Bitcoin payments faster and cheaper.

Metaverse: A virtual network of spaces where creators, gamers and AI can collaborate. Think of it as one big playground for users to build, socialize and even earn crypto.

Miners: Actors in a blockchain network that can earn the right to verify transactions and add new (valid) blocks to the chain.

Network Effect(s): The phenomenon where a product, good, concept or service increases its fluidity as it becomes more valuable for each user, producing compounding value or worth.

NFT (Non-Fungible Token): NFTs are assets that live on the blockchain; they are unique and can be used to show ownership or validity.

Non-Fungible: Non-Interchangeable (see definition for NFT).

Permissionless: Not requiring authorization.

Protocol: The rules that govern a given technology; they determine how to format, process and transmit data.

Proof-of-work (PoW): A consensus mechanism where miners use computing power to secure a crypto network and keep it running. This effort is *proven* to the network through cryptographic functions, allowing miners to verify transactions on a blockchain.

Satoshi Nakamoto: A pseudonym for the creator(s) of Bitcoin.

Smart Contracts: Pieces of computer code programmed to produce a predetermined outcome that allow us to automate events, actions and results; a trustless agreement or promise made digital.

Web 3.0: An evolutionary stage of the internet, where decentralization, digitally native money, autonomous systems, and self-sovereign identity combine to enable new economic models.

White Paper: A thesis paper, such as the one written by Satoshi Nakamoto outlining the Bitcoin protocol and its use as a "peer-to-peer electronic cash system" and "distributed timestamp server." Don't worry, you have this whole book to get caught up…

What is a Bull?

Throughout history, the image of the bull has been used to capture a key sentiment in finance (and now crypto) towards the market, or specific assets. This attitude is one of hope.

Representative of an investor's gutsy outlook, the bull defies odds and manifests action. It demands a rosy outlook, despite potential signs suggesting otherwise.

Wall Street stock or...well, livestock, this symbol traditionally occupies a masculine role. The bull is a male, known for its sustained energy that shows no signs of softening.

Horns poised for achievement, ready for the ride — there's a lot about this archetype that reminds me of *women* and the strength we bring to every endeavor in life.

The women in my life are bulls in the best of ways. Each of them has weathered ups and downs, taken a warrior stance when the odds are stacked against them.

Growing up, I disliked money — anything to do with it: wealth was all around me, and it instilled fear within my core. I didn't want to be selfish, so why would I ever want to learn about *money*? I swore I'd never work in a male-driven field with earning at its core.

In my twenties and thirties, something shifted. My approach was unsustainable, and I was sick of relying on other people for (you guessed it) money. For years, I looked for my independence everywhere. In learning about Bitcoin, cryptography and the promise of decentralized technology...I grew horns.

For the first time in a long while, perhaps since the early days of the internet, technology felt like a means of freedom. A way to grow. A source of inspiration. I wanted to learn everything about this space, to soak it all in — and to invite everyone I knew on board, despite the rocky road underfoot.

Meeting Maria in The Crypto Academy was the best thing that happened to my career; but more than that, it gave me an outlet to be completely bullish on life and the challenges that lay ahead.

Whether you feel bullish most days, or find yourself more comfortable in the cave — maybe you're already crazed about crypto or utterly bearish about it — I hope this collection of thoughts and resources can give you a little bit of courage to dust off your boots...and feel *hopeful* that it's never too late to learn something new.

Stay Wild,
Stevie Sats

Introduction:
A Note from Two Women in Crypto

This collection of email newsletters was written over the span of nearly one year, from late 2021 to 2022, for the purpose of education and community building.

From the start, our aim was to empower women from all backgrounds and experiences to flourish in crypto and Web 3.0 — whether that required stepping stones to a new career or fluency in a room full of fintech bros. As soon as we started feeling confident, it became our mission to share the love.

Together we began to write about our favorite crypto topics through a Substack newsletter called Womxn In Crypto. We took great joy in watching this community take shape and grow — gaining perspectives from all different women working on projects in the Bitcoin and larger crypto space.

Now, in this second chapter of our crypto journey, our goal is to distribute this work through alternative mediums — so that readers have access to our thoughts and discoveries on crypto no matter where or how they like to read.

The majority of this work is the same as it first appeared in our newsletter online, but it has been restructured so that it's easier to navigate via eReader or in print. Content has been edited and added for further clarity (and to represent any significant evolution in the cryptocurrency space since the original pieces were written).

Our mission remains the same:

Womxn In Crypto is a newsletter for women who want to learn through straightforward and accessible stories. We help you understand the complex world of Bitcoin and the spectrum of decentralized digital currencies so that we can all succeed together.

These chapters were individually written, with careful peer review from both sides on each topic. You may notice some variation in tone and writing style throughout the book; this reflects a shift in who is writing the piece. We wanted to maintain our authentic styles because that's what makes us...us! But we've included some transitions and consistent language so you know it's a package deal.

A small number of articles were first published on outside platform HackerNoon by Stevie Sats and later included within this collection for an editorial perspective on specific topics. This selection includes "Edgar Allan Poe Was Bullish on Crypto," "Non-Fungible Sentiment," "Deja Vu with Goggles On" and "Decentralized Metaverse and the VR Spectrum."

Crypto is always changing, and there's never a dull moment in this space. We've selected and adjusted pieces that we feel best reflect the space currently — while attempting to paint a realistic picture of what's to come. Surely, there will be developments ahead that may alter our position or understanding on several topics — but our eagerness for financial revolution, and the promise of women driving this change, is something that will continue to shape our worldview.

In the chapters that follow, we'll discuss our entry into the world of crypto and what we've loved learning about most in our early studies. We'll lead with Bitcoin and the foundations of this world-changing tech—but we'll also share plenty of information on altcoin developments, like the Ethereum network and DeFi, too.

We hope you love it. We hope that you'll form your own unique perspective on cryptocurrency, while teaching others. This journey has launched our careers and changed our lives, and we're delighted to share some of our best tools with all the women who make this world an incredible place to learn.

Thank you,
Maria and Stevie

PART I

Bitcoin: In Case You Missed the Gold Rush

THREE THINGS I WISH I KNEW BEFORE GETTING INTO CRYPTO

For some, the road into the world of cryptocurrency is slow and smooth—accelerating steadily over time, perhaps starting with a sole focus on Bitcoin while learning more about its past and present along the way. We could liken it to Scrooge's journey in Dickens' *A Christmas Carol*: fiscal frustration followed by sound guidance into a deeper appreciation of community and coin alike.

For others, it's a whirlwind of learning, unlearning, and investing all at once. There's often self-doubt, market FOMO and plenty of misinformation along the way. This is how I felt getting into crypto. If the aforementioned learning experience mirrors Charles Dickens, then mine was more like the movie *A Christmas Story* — it felt like I was given a Red Ryder BB gun with no instruction manual.

This was thrilling, even for a pacifist, and surely some conversations felt like I was walking into a room full of people while wearing pink bunny pajamas. I was confused and concerned.

In some ways, though, this plunge has been for the best — enrolling in classes for the first time in years, seeing the world through lighthearted tweets and memes, scouring newsletters during one of the most eventful years in the history of fintech. NFTs and laser eyes exploded onto the scene while I scrambled to understand what a "golden cross" was. I fell in love with all of it.

Arriving later to the party (or at least after cocktail hour) had its ups and downs. Mainly, there are a few core principles that would have been nice to know before becoming fully entrenched in crypto. I hope that by writing about them, I can provide you with some of the insight I wish I'd had sooner. Whether you're an old pro or a novice, it might help to familiarize yourself with these three mantras:

1. Not Your Keys, Not Your Coins.

Custody is *everything* in crypto. I entered the industry when public exchanges were already established. Even though I knew that public and private keys were the essential element of Bitcoin ownership, I questioned the reliability of memory. Centralized exchanges like Coinbase and Gemini seemed to conveniently take care of that for me.

What I came to realize: There was more to it than convenience.

I was taught the value of self-custody and introduced to ownership options (from software wallets to hardware ones like Ledger and Trezor) in online courses, where I began to ask myself what solutions would make the most sense for me long term. Lifestyle and risk tolerance both play a part — and it all came down to where I felt comfortable placing my trust. Most importantly, I learned this: if you don't have the password for your crypto, that means another entity technically owns it.

2. Don't Trust. Verify.

This saying is often used in blockchain and crypto communities to counter the old Reagan-era adage,[1] "Trust, but verify." This attitude was used frequently to speak to compliance between the U.S. and the Soviet Union through dual-sided confirmation. Bitcoin enthusiasts have flipped this phrasing on its head to instead point out that peer-to-peer trumps trust. In a trust*less* system where every transaction is verified by individuals within the network, there's no need for a third party to authenticate.

Of course, there's a nod here to the verbiage on the U.S. dollar: In God We Trust. Faith in government-printed currency, in this sense, also begs to be examined. Any proposed solution to the blunders of our financial system should aim to address, course correct, and continually innovate.

3. Not All Coins Are Created Equal.

For better or worse, what started with an anonymous white paper has sprouted a bustling crypto ecosystem consisting of many currencies across a spectrum of decentralization. Altcoins — tokens other than Bitcoin — are scrutinized for their levels of "decentralization" (even Ethereum), as they are all built differently and each bring something completely different to the table.

When something is decentralized, that means it can operate without relying on a central authority or nucleus of control. It's argued that Bitcoin is the only asset with a protocol that doesn't yield to some central authority, a feature secured by its proof-of-work consensus mechanism (a term we'll break down in the following chapters).

Much of my first year of growth was spent researching "both

sides of the coin" — to better understand the structural issues that Bitcoin solves for before blindly championing any coin that might falsely claim to do the same.

When I was ready to find my place along the Bitcoin adoption curve, clues like these helped me to find my footing in fintech and improve my fluency in crypto.

I'm pretty sure many of those who got in the game years ago encountered similar turning points, and perhaps an even rockier emotional ride.

No matter when you start, the crypto journey is all about long-term learning. But for those who are just starting to ask questions now, it's a good thing there are so many resources to explore — from independent news to podcasts and events. Asking questions is the most important part, and as long as you're doing that, your growing knowledge may eclipse many smaller regrets.

NOTE: According to IMDB, *A Christmas Story* was based on the book *In God We Trust, All Others Pay Cash*.[2] Loving the irony here.

* * *

WHAT IS BITCOIN?

Bitcoin is much talked about but little understood. I believe it will continue to grow in significance as more women realize the benefits it can bring as an investment opportunity and as a key weapon in the battle for financial freedom.

In this chapter, I want to answer the question most women ask when they first encounter Bitcoin — *what is it*?

Once we've covered that, I'll explore some key aspects of Bitcoin that complement this introduction.

The Bitcoin network is a protocol that facilitates the digital transfer of value without the need for a trusted intermediary. It uses blockchain technology to validate transactions. Bitcoin is the reward given to the validators, often referred to as miners. This reward is given in the form of digital coins — bitcoins.

The digital part of that description means that Bitcoin has no physical form to touch or hold, unlike a bank note.

A good way to think of this is to compare it to the U.S. dollars that exist in a bank account. If every person, company and government wanted physical notes in exchange for the money in their bank accounts, it wouldn't be possible. That's because most money only exists in digital form today.

Taking this a step further, a key difference between the digital money that exists in a bank account and the digital assets that exist within a blockchain network is that digital money is controlled and managed by a bank. The bank stores the information in a database of some kind, which exists on a network of computers that the Federal Reserve uses to communicate with the banks.

Bitcoin is different. In this case, the information (about each bitcoin transaction) exists on a distributed ledger within a decentralized blockchain network. If you're unfamiliar with these terms, that's understandable. You are most likely to encounter them in finance, accounting and computer science — but more and more, they're starting to show up in everyday language.

To understand the distributed ledger, we can compare it to the bank example referenced earlier. Whereas the bank has a centralized database recording the money it controls on behalf of its customers, Bitcoin's distributed ledger is essentially a transaction history that is shared by a network of computers around the world. The blockchain part of the description refers to how the data is stored, in a series of blocks that are linked one after the other in a chain, each containing a hash of the data from before (linking to the previous block).

This network of computers is referred to as the Bitcoin network and it is decentralized because there is no single central point of control — or attack. The bitcoin on the distributed ledger belongs to addresses that can be accessed using public/private key cryptography (hence the name "cryptocurrencies"), and every computer on the network has a copy of the Bitcoin ledger.

This ledger records the number of bitcoins associated with each address. The transferring of bitcoins between addresses is managed through a "Proof-of-Work" consensus algorithm. You can think of a consensus algorithm sort of like a "terms of agreement" among the users of a network that's baked right into the code. In the case of Bitcoin, computational energy is used to verify transactions on the ledger, keeping it decentralized.

Simply put: the work that the miners put in, via computation, is what allows Bitcoin to operate securely — free from the interference or oversight of corporations, governments and central banking systems.

In following chapters, we'll discuss Bitcoin as a store of value and a means of exchange. These two roles are often seen as criteria for what makes money...well, sound money. As we've mentioned, bitcoin is seen as a type of currency, but it's also more than that. Bitcoin is a network that serves as the scaffolding, the support system, that provides the ideal environment for this currency to thrive.

We can use an analogy to better understand this concept: just as gold was used to back the U.S. dollar (under the gold standard) until 1971, the Bitcoin *network* is used to back bitcoin as a currency. This reasoning should not imply that Bitcoin and gold are the same (they have vast differences, which we will address in an upcoming chapter). However, this sentiment can help us understand the twofold nature of this innovation and how Bitcoin supports itself instead of relying on any third party.

Though shrouded in secrecy, the paradox is overt:

Bitcoin is a representation of value that proves its inherent worth over time.

* * *

TOP FIVE TRUTHS ABOUT BITCOIN AND HOW IT WORKS

In the previous chapter, I briefly answered the question "What Is Bitcoin?" and explained some of its main features. The next natural question is "Why should I care?"

Let's explore a list of five Bitcoin truths that answer this question.

1. Bitcoin is self-sovereign money that can't be stopped

Bitcoin is self-sovereign money because you control it. This differs from the money you hold in a bank account, which is controlled by the bank — and ultimately the government. The bank could take away your access by freezing your account, especially if the government told it to. Bitcoin is different because only you have access to your bitcoin via your own private keys.

This may not seem important if you live in a First World country, have a regular job and have never been in trouble with the law. Unfortunately, for many brave female activists around the world, this is not the case — and the self-sovereign nature of Bitcoin has helped them to retain their assets or pay for their struggle, even when their government opposes what they do.

2. Bitcoin is a scarce asset in a world of free money

The fact that there will only ever be 21 million bitcoins matters deeply because scarce assets look increasingly attractive right now.

Ever since the financial crisis of 2008 and particularly in response to the COVID pandemic, governments around the world have printed huge amounts of new money.[3] Whether you agree with this policy or not, it is hard to disagree with the view that massive amounts of extra money in the system leads to asset price inflation.

The fact that there will only ever be 21 million bitcoins is important because, as more money is created, more investors may see the value of all fiat currency decreasing and choose scarce assets that they believe will retain value.

3. Bitcoin is a store of value over the long term

The title of Part I mentions gold for a reason. Bitcoin is compared to gold by some investors because both are scarce assets. Gold is widely considered to be a good store of value, but many gold investors would reject the idea that Bitcoin should also be thought of in this way. This distinction is grounds for fierce debate.

It is certainly true that Bitcoin's price can fluctuate dramatically (this is referred to as *volatility*), as it did in 2021 when it dropped over 50%. However, this doesn't mean it hasn't been a good store of value over the long term. Many economists and scholars make a case for this, arguing that Bitcoin's scarcity, divisibility, portability and fungibility — features we'll dive into in the next chapter — could make it a more reliable store of wealth and potential hedge against inflation than traditional currencies, as we see the latter become devalued over time.

Bitcoin is an emerging asset class: this innovation behaves differently than traditional stocks and bonds, and it offers value through traits we're not used to seeing in the market.

As newcomers adopt the tech, the general public will likely become more comfortable with it.

But in the meantime, investors and hopefuls ride out the waves as its early years see tremendous volatility. Individuals who bought Bitcoin at the peak of the 2017 rally saw the price rise again years later, even though the price fell sharply in the intervening period. And on a day-to-day basis, there's fluctuation to consider. The volatile moments may signal hope or frustration depending on a person's investment style and time preference (day trading versus dollar cost averaging, for example, are two different but valid approaches to unpredictable market conditions).

In sum, Bitcoin has its ups and its downs. But many analysts believe that over time, the volatility may settle as it becomes more widely adopted. And because of the value in its design and purpose, many also believe that it's actually a *more* secure asset to place their wealth in compared to fiat currency (which loses its value), or gold (which doesn't carry as many of those desirable attributes that we listed above).

4. Bitcoin is the most battle-tested cryptocurrency

The first bitcoin was mined at the start of 2009, making this monetary network over 14 years old now. In that time, the network has never been hacked. However, many people believe that Bitcoin has been hacked, either because of misreporting in the press or because they confuse the Bitcoin network with applications that sit on top of and interface with the network (e.g., cryptocurrency exchanges) that *have* been hacked.

The fact that the Bitcoin network has never been hacked matters because it seems almost certain that criminals and hostile government agencies will have tried. There's no guarantee Bitcoin won't be hacked in the future, but this track record should reassure anyone concerned about how secure an investment it is.

5. Bitcoin does use energy (and that's not a bad thing)

There has been a lot of coverage about the energy the Bitcoin network uses, and we feel it's important to examine this topic. Energy is used in the process of mining to keep the network secure.

Without going into too much detail, energy is used by miners in the proof-of-work consensus mechanism. This involves them racing to solve complex puzzles and using large amounts of computing power to do so. The miner that solves the problem first gets the opportunity to add transactions in a new block and receives bitcoins as a reward for doing so.

What are those complex puzzles, you ask? Cryptographic ones

— and we'll tell you all about them in an upcoming chapter.

The fact that this process requires a lot of energy plays an important role in making the Bitcoin network secure. So, if you believe having a secure, digitally native currency that cannot be devalued at the press of a button by a central bank is important, you will also likely agree that Bitcoin's energy footprint is worthwhile. The Bitcoin network also uses electricity at the margin, i.e., where its cost is lowest, and at times it is not used by "typical" power users. Because it is difficult to "store" off-peak energy that is otherwise wasted, the Bitcoin network serves to provide an economic return for electricity providers — while helping to drive the clean energy revolution (as that is ultimately a cheaper form of energy).

If you are interested in diving deeper into this topic I would encourage you to read the article "Bitcoin's Energy Usage Isn't a Problem. Here's Why," by Lyn Alden.[4] You can find it on her website lynalden.com, or many other podcast and audio streaming platforms.

* * *

BITCOIN IS A PROBLEM SOLVER

Bitcoin is widely known for its role in the crypto and financial ecosystems as a new asset with lots of volatility and grounds for regulatory debate. In many ways, Bitcoin is a disruptor.

That's not necessarily a negative. Most innovative tech that sees some sort of long-term success has to, in some way, disrupt the entire space to be productive. Yet beyond the feisty price action and topsy-turvy headlines, it's worth taking a look at what else this crypto pioneer does differently — what makes it so unique in the first place. Those are the attributes that allow it to shake up the system and give it sticking power.

On a Protocol Level

Have we already mentioned what a protocol is? Who's up for a refresher?

You can think of a protocol like a set of rules that governs a network or system, and how information (data) is kept, or used or communicated. We'll loop back to this in a minute.

Embedded within the Bitcoin protocol is a very special trait. In fact, this nifty feature is what sets it apart from earlier attempts at digital currency (yes, there were others! Hashcash, for example, was proposed by cypherpunk Adam Back before Bitcoin entered the scene — and advocates of privacy and cryptography have been attempting to make internet money a thing since the '90s).

Bitcoin solved something called the double spend problem,

which means any entity spending the same funds twice (aka cheating the system).

Why is this "double spend" such a problem? You can think of it like trying to sneak a friend into a movie theater. Let's say the movie theater can only accommodate twenty viewers per screening, and you and your friend are the first two in a line of twenty-one people. You plan to pass your ticket to your friend after being let in (without getting caught) so that she doesn't have to pay for one herself.

The theatre employee, or ticket scanner, checks your ticket briefly and lets you in. Assuming there's no way of knowing the ticket is *the same one*, your friend will also be able to view the film with your ticket. After twenty people, the theater is full. The person at the end of the line is allowed in, too, but searches around frantically for a seat. You've taken a comfy spot in the second row, and that person from the back of the line gets cheated out of their hard-earned money: they have to stand for the entirety of the show.

This example isn't exactly the same as blockchain, but the idea's similar: if all tickets (or coins or tokens) are seen as legitimate, the real ones lose their value.

Bitcoin disincentivizes, and essentially nullifies, the ability to double spend. Satoshi Nakamoto outlined how to achieve this in his white paper for the new tech.

To understand this, we need to get familiar with hash functions. Hash functions produce a unique string of data assigned to each batch of transactions in the network as it gets approved. Like a Dewey decimal number for each monetary exchange, or a unique fingerprint for a human being.

Back to the movie theater example. Say the ticket scanner checks the barcode of each ticket, which corresponds to a number in the system. If someone tries to submit an already used ticket, or an old one, the number will register as invalid — and the scanner will notify an employee. They are then able to politely turn the fraudulent customer away, and only allow in the twenty people who paid for their tickets.

This analogy can be extended to blockchain technology. If a person wants to copy a transaction and send it out a second time, the system has a way of knowing it happened — because timestamps have been assigned to each transaction batch. The *hash* will match the transaction that has already been seen or acknowledged (similar to a movie ticket that's already been scanned), and so it becomes verified in this system.

This process is the same one that also creates new bitcoins and circulates them into the supply; as miners process each transaction, they are rewarded with bitcoin for doing so. We have arrived back at the concept we defined earlier as proof-of-work consensus.

Since the Bitcoin blockchain is a decentralized ledger that records exchanges that are sent and received, and it's viewable to the public, every computer in the network has an updated copy of the information. The server then broadcasts the hash to all the computers in the network, so it's fully visible.

Right here, we have perhaps the first element that makes Bitcoin a true problem solver. Hash functions and the solution to the double spend problem form the foundation for all that cryptographic puzzle-sleuthing stuff that comes along with mining. Altogether, it's safe to say Bitcoin was born to provide solutions.

On a Value Level

In the last chapter, we mentioned a few fancy terms: scarcity, divisibility, portability and fungibility. We also talked about Bitcoin as a store of wealth and a potential hedge against inflation. All of these are traits that go to bat against certain financial problems or undesirable outcomes.

Let's get into what these words and phrases actually mean and why they're significant. But first, it's necessary to understand what gives something *value*.

Value has been influenced and determined by various factors since the dawn of society. Whether bartering and trading goods and services or forming social hierarchies, groups of people appraise worth over time — giving it meaning.

One group might recognize a natural object (such as gold) as having worth based on traits like scarcity. They could also mint their own form of currency using more common resources (like paper money). Some people place a higher price still on utility: how useful an object, service or resource proves to be (perhaps land for homes or carpentry).

In other words, perception can change throughout time, but as long as this worth is agreed upon (by a community, tribe or society), that's what makes it valuable.

Here are some of the many attributes that give Bitcoin perceived (and therefore actual) value.

Scarcity: We explained in a previous chapter that Bitcoin caps its currency supply at 21 million. Because there's a limit to the amount of bitcoin that can ever be created, it can't be over-printed, like fiat dollars can. Bitcoin is scarce. This over-printing of fiat currencies creates inflation: a decrease in the value of a currency over time. So you'll often hear people refer to Bitcoin as a hedge (boundary or protection) against inflation for this reason.

Divisibility: It is not very practical to subdivide solid bars of gold accurately, or to divide fiat currencies smaller than their smallest denomination, such as one cent. This can be a problem if you want to buy or sell things worth less than one cent (microtransactions) or if you want to sell only part of your gold bar. Because bitcoins exist digitally, they can easily be split into very small subdivisions. Small Bitcoin transactions are sometimes denominated in satoshis (named after Satoshi Nakamoto and abbreviated as "sats"), where there are 100 million satoshis in a bitcoin.

Portability: Because Bitcoin is digital, it doesn't need to be lugged around like stacks of printed dollars or heavy gold bars...or any other physical object that a group of people deem to have value.

Fungibility: The ability of an asset, commodity or good to be exchanged or replaced with something of equal or greater value. 1 BTC = 1 BTC.

Store of Wealth: A place to park your funds, with the aim that they may keep their value or increase in value.

Let's add in a few more solutions that Bitcoin proposes as an intended global currency...

Immutability: Unable to be changed. Bitcoin's blockchain operates in this way: a block cannot be removed or changed once it's added to the chain. Transactions are stored as is and can't be reversed or edited. This makes the network safer and more secure, as no one can go back and try to alter the blockchain's history.

When we think about making changes to Bitcoin's source code, it's important to remember this idea. If developers were to decide to change the code, then something called a "fork" would happen, which would result in the creation of a new coin/network entirely — one that wouldn't be Bitcoin anymore.

Auditability: When data can be measured or reviewed for factual accuracy or fairness. In this case, a blockchain is auditable because by its very nature its transactions can be viewed and analyzed. The blockchain is a public ledger of recorded data that's available for everyone to access.

Anonymous/Pseudonymous: A key feature of the Bitcoin network, and motive for its creation, was for this technology to provide levels of anonymity. While we can trace any transaction to its wallet address (where the bitcoins are sent to/from), the sender/receiver can remain unknown.

This gets a little more nuanced with KYC (Know Your Customer) identity verification, and can also be explored when learning about privacy altcoins, which we won't detail here. Just know that it's something we suggest you research if you want to learn more about privacy and anonymity.

Permissionlessness: No authorization is required for the Bitcoin network to operate or to make and approve transactions. In other words, Bitcoin does not require a bank or a government, or any third party.

Let that one sink in for a little while.

Now back to that word we mentioned earlier: protocol. With centralized structures, we typically have a third party running the show. But if a protocol, the code, is doing the governing — as it happens in the case of Bitcoin — then we could say the *code is law*.

In this way, Bitcoin doesn't require the permission of a person or group or entity because the code allows it to operate independently. It's a network, it's currency, and it's also completely sovereign.

Bitcoin has all of these traits (and more!), and they are all solutions to problems surrounding cash and other measurements of value.

On a Global Level

Earlier in my crypto learning journey, I adopted a too-narrow scope. I tried to keep tabs on news and policies in the U.S., but I made little effort to understand its impact on the world at large. I made assumptions about digital finance that were informed by my American upbringing, niche surroundings or local political climate. This is the worst approach to *any* kind of education.

Since then, I've opened my worldview. Learning alongside peers from El Salvador, Argentina, the U.K. and Asia has helped me understand more use cases for crypto and learn from events that have happened outside my home country. Writing these newsletters together has also helped us grow in this way — as two women with completely separate, nuanced histories. We grew up in different countries, experienced different challenges — but we connected on our passion for education and what the future of money might look like *everywhere*.

Part of the beauty of Bitcoin, in particular, is that it is a global digital currency; it has the power to build peaceful, equitable bridges and connect individuals around the world. The surprises that come with shifting to this global mindset are endless.

Bitcoin is often seen as a great socio-political equalizer: It's a money that doesn't care about — or even know — who you are, what you do, where you live, or what you believe in.

We mentioned inflation earlier, and while it's seen as an inconvenience to some, there are many people around the world living in societies where it's a hugely immediate issue. Their currencies may be so debased that it's difficult to support the nation's infrastructure or individuals' and families' lives. A scarce commodity or good, then, may be looked to as a replacement.

Some even say that Bitcoin promotes peace, as a non-violent money. Many argue that its non-physical nature makes people less likely to hurt or even kill one another over it, since pointing a weapon at a person to burglarize their home for gold or cash may be a likelier and easier feat than trying to take down a system that disincentivizes its attackers, simply by virtue of the way the code was built.

There's a lot to unpack there, since it leads into a much larger discussion on blockchain attacks and control...but worth bringing up.

A new notion of what money is, and how it's owned or transferred or carried, sets up an entirely different framework around how people might seek to usurp it. So at the very least, this requires that we consider it as something entirely different from money as we currently know it.

* * *

WHAT MAKES A NETWORK GOOD

"We know that software can't be destroyed and that a widely dispersed system can't be shut down." —Eric Hughes, "A Cypherpunk's Manifesto"

The growth of social networks has been the most influential change in our society over the last decade. It can often feel like these tools have been part of our culture for much longer than that; many of us can't imagine our lives without them. For the first time since the dawn of the internet, social media has allowed the network effect to take hold globally. Because of how strong these connections are, they can shape grassroots movements and humanitarian impact in incredible ways. But there are other players influencing the game.

We know that Facebook and similar companies have hooked us on a product — one that functions like a Matryoshka doll of other products. This platter of "choices" is connected by people and companies, and their power lies in a web of trust and interest.

Digital platforms facilitate interactions or exchanges, and their architecture is a nucleus of ongoing innovation. It's smart for businesses to model themselves in such a way that the network becomes the machine; the service, data and audience are all built in. Until this innovative design becomes, as VR pioneer Jaron Lanier would put it, "a manipulation engine."[5]

Social media is a heady combination of advertising, addiction principles and influence. Now, what if monetary transactions were the social tool? They would have to be free from each of these traits and sit very much outside of any central person, institution or system. There may always be bad actors...but if those actors don't have control, there's nothing for them to gain.

When I first learned about Bitcoin, I recognized that its value was more than the sum of its parts. By providing people with an autonomous and meaningful use of currency (versus any that paper money arbitrarily assigns), it proves itself more and more as a store of value for humanity as users adopt it.

The last part of this concept has a name: the network effect. It refers to compounding worth or benefit and how users of a community, product or other good determine and share value. The network effect is touted as one of cryptocurrency's most successful attributes; it's how we come to acknowledge and rely on a tool that's inherently useful and self-sufficient.

One illustration of this impact lies in Bitcoin's layer-two Lightning Network — where quick transactions of satoshis can take place instantly via smart contracts. If more and more users find significance in this updated digital process, it can increase the value of Bitcoin and promote it organically amongst masses of people based on utility alone. We've seen indications of this type of growth before — for example, when the number of nodes on the Lightning Network doubled in a mere matter of three months.[6]

This is exactly where Bitcoin's network effect deviates so sharply from that of Facebook or Google, for example. There's no advertising machine, no corporate incentives and no business model. As a currency, its protocol is immutable and permissionless. It functions without a third party overseeing, tracking or authenticating — and as we know, third parties so often have interests.

Yes, I just compared a monetary system to social media empires, and they could not be more different in purpose and function — but humor me for a moment. The overlap lies in the fact that they are networks poised to grow.

Unlike networks that use algorithms or advertising to make choices for the individual, a cryptocurrency network could run on autonomy and encourage financial freedom. Bitcoin has shown us that blockchain technology has the potential to function as a framework for equality. As a peer-to-peer payment system and value supply, this invention could help level the financial playing field and encourage communities to implement supportive tools and ongoing education — without discriminating on the basis of race, gender, background or beliefs. These are a few positive potentials of decentralization, but they will only work with a willingness to adjust old systems.

Of course, some will speculate that Bitcoin's network effect is well intentioned but could veer off track, if investors put all their eggs in one basket and bank on the benefits of rare outcomes. It's also important to maintain awareness of any one entity using the biggest slice of the pie to influence others (and note that this entity could be anonymous). We should consider all aspects of networks and their implications, so that we can steer digital currency towards its original aim: globally sound solutions rather than a repetition of prior mistakes.

In our day-to-day world, there are lots of healthy, organic networks out there. Families. Friends. Free-of-cost educational programs. Maybe your local yoga studio or book club is a valuable network to you. Community gardens are one of the best examples I can think of when it comes to sharing profit or harvest — a thriving ecosystem (kickstarted by humans) that builds upon itself into the future, benefitting more and more people each year.

It takes conscious effort to pry oneself away from the influence of draining networks. There are social platforms I feel obligated to use for business, and others that help me stay connected to family and friends. By joining, creating and supporting constructive and autonomous networks, perhaps more of our connections in coming years will be sources of reciprocity and pride.

NOTE: There's a world of information available for those

interested in learning more about organizations and causes devoted to privacy and digital rights. Do a little digging, and you may just find some answers (or open up new inquiries) into this important theme — beyond its relevance to crypto-currency. We recommend "A Cypherpunk's Manifesto," referenced at the top of this chapter, as a starting point.

* * *

WHAT IS THE LIGHTNING NETWORK?

The Lightning Network is a second-layer solution that aims to make Bitcoin payments faster and cheaper.

It is an important part of the Bitcoin community's plan for increasing the utility of Bitcoin as a new monetary network and is therefore a key development you should understand.

Why Does the Lightning Network Matter?

One of the main criticisms of Bitcoin is that it has failed to live up to the original vision of 'a peer-to-peer electronic cash system' that was set out in the Bitcoin whitepaper.[7]

This criticism revolves around payments being too expensive, which is because Bitcoin transactions are confirmed by being entered into new blocks that can only hold a limited number of transactions.

The Lightning Network was proposed as a secondary layer that would enable the volume of transactions to scale while keeping the base Bitcoin blockchain secure and decentralized.

How the Lightning Network Works

The Lightning Network works by establishing one-to-one payment channels between users and then scaling this model so that multiple addresses are connected.

On the simplest scale, two users (who may be referred to as nodes) send an amount of bitcoin to a shared multi-signature address (one that requires multiple signatures, rather than just one, to authorize each transaction) and this opens a payment channel between them. They can then send smaller amounts of Bitcoin to each other, as long as these amounts don't exceed the shared total.

These transactions are recorded on the Lightning Network but not on the Bitcoin blockchain. Only the opening or closing of the channel is recorded on the Bitcoin blockchain. Thus, the Lightning Network allows users to avoid the fees associated with Bitcoin transactions.

The Current State of The Lightning Network

As more payment channels are established, there are more individual users who have one-to-one connections with multiple users. These users then start to act as routers. As a result, a user can pay another user they are not directly connected to simply because they are both separately connected to a routing user.

According to 1ML, a Lightning Network search and analysis engine,[8] there were around 35,000 nodes and around 87,000 channels in the Lightning Network in early March 2022. This translated into a network capacity of around 3,500 BTC or roughly $135,000,000.

Use Cases for the Lightning Network

A number of projects, including those from private companies and national governments, show how the Lightning Network is being used right now.

Strike app

Strike is a payment application built on the Lightning Network that allows users to make payments at virtually no cost.

Twitter tips

Twitter has enabled tips, which are micropayments made through the social network, using Stripe and the Lightning Network.

Nostr zaps

Nostr is a decentralized protocol that can be used by developers and content creators in similar ways as traditional social media, only without the guardrails or high levels of censorship.

This infrastructure complements the Bitcoin ethos by offering an instrument for liberated speech — along with opportunities for Bitcoin monetization on the platform. The term "Zap" is used across Nostr clients (apps or platforms) for receipts displaying Lightning integration for sending satoshis back and forth.

El Salvador

This country has made bitcoin legal tender and has introduced Lightning Network integration for its citizens.

* * *

UNDERSTANDING BITCOIN SCALABILITY

Bitcoin scalability refers to whether Bitcoin will be able to handle the large number of transactions necessary for it to compete with existing payment networks.

In the previous chapter outlining the Lightning Network and its role in digital payments, we described what is probably the most established Bitcoin scalability solution. However, there are other options, as well as an important backstory that is worth knowing.

The Story of Bitcoin Scalability So Far

Even though the full title of the Bitcoin whitepaper is "Bitcoin: A Peer-to-Peer Electronic Cash System," most people think of Bitcoin as more analogous to digital gold than digital cash.

One of the reasons is it cannot currently handle a large number of transactions at the speed required for it to work like cash. This is because the number of transactions that can be added to a Bitcoin block and the time it takes for a block to be mined both restrict the network's scalability.

There have been attempts in the past to change these to increase the number of transactions that can be processed. One of the most famous resulted in a hard fork of the Bitcoin blockchain, which became Bitcoin Cash and was subsequently forked again into Bitcoin SV.

However, the sustained growth of the Bitcoin network shows how most have accepted the limits of Bitcoin's scalability in order to keep it secure and decentralized.

The Current Bitcoin Scaling Landscape

Although these attempts were rejected by most of the Bitcoin community, that doesn't mean it has given up on trying to scale the network by other means.

The Lightning Network

The Lightning Network is a layer-two solution that aims to increase scalability by letting users establish one-to-one payment channels that are separate from the Bitcoin blockchain. As these one-to-one channels increase, individuals can start to send payments to those they are not directly connected to, via other users who are connected to both of them.

The Liquid Network

The Liquid Network is slightly different in that it is a sidechain rather than a layer-two solution. This means it is a unique blockchain with its own Liquid Bitcoin or L-BTC tokens that are one-to-one exchangeable with bitcoins. Since one Liquid block is processed per minute (as opposed to Bitcoin's 10-minute block time), this solution allows for increased transaction speed. However, it's worth noting that the Liquid Network is more suited to moving many bitcoins quickly and privately than it is to enabling small payments.

Will Bitcoin Be Able to Scale?

While it's always difficult to say this with certainty, the signs are looking good for Bitcoin scaling. Both the Lightning Network and the Liquid Network are growing gradually, with various entities around the world starting to use them and no major issues having occurred so far.

* * *

A SIMPLE GUIDE TO BUYING BITCOIN

Investing in Bitcoin (BTC) might be one of your crypto goals. For convenience's sake, we have broken this down into four simple steps. In previous chapters we covered "What is Bitcoin?" and "Top Five Truths About Bitcoin and How It Works." This guide will build on that foundation with a few practical insights on exchanges, payment, storage and more.

1. Choose Your On-Ramp

The term "on-ramp" describes a vehicle, platform or product that allows the user to swap their fiat (traditional) money for crypto.

Cryptocurrency exchanges have historically been seen as the most convenient option for many, since they offer a large variety of features and a list of cryptocurrencies for trading. They're often referred to as "fiat on-ramps," since they offer an opportunity to exchange traditional (fiat) currency into bitcoin or another crypto. The user experience might feel akin to familiar financial trading vehicles, and they're usually owned by centralized companies.

Cryptocurrency exchanges will allow individuals to buy, sell and hold cryptocurrency. A known best practice is to select an exchange that allows users to withdraw crypto to their own personal wallet.

There are many types of cryptocurrency exchanges, and the most popular exchanges are the centralized ones that follow laws requiring users to submit identifying (KYC — Know Your Customer) documentation.

Keep in mind that some exchanges have collapsed, whether it be the fault of bad actors or poor infrastructure. It is important to do your own due diligence, and to understand that there is significant volatility in this space; anyone may lose the entirety of their crypto investment at any time.

Centralized exchanges, though often accessible, aren't the only method of attaining Bitcoin or other cryptos. On-ramp options also include:

- Over-the-counter trades (directly between two parties or facilitated by another party)

- Decentralized exchanges, aka DEXs (a

peer-to-peer, non-custodial option that facilitates exchanges of crypto)

As with all of decentralized finance, and crypto generally, there are risks. DEXs are typically aligned with more seasoned traders and high risk-tolerant investors due to the steep learning curve. Centralized exchanges carry different kinds of risk in their role as custodian, since you might rely on them to hold your keys. Either type of exchange should be considered for its unique traits and the responsibilities that come with each.

- NFT marketplaces (more on this in an upcoming chapter)

- Airdrops (promotional distribution of digital assets like NFTs or cryptocurrency, usually for free, to attract interest or buzz around a project)

- Gifts (sent from/to different Lightning wallets, for example)

- Mining (By setting up a Bitcoin mining rig and joining a mining pool, you can earn cryptocurrency as a reward for adding new blocks to the blockchain)

2. Verify and Connect Your Exchange to a Payment Option

If choosing an exchange, the user will need to gather personal documents (e.g., driver's license, social security number, etc.) and submit them to the exchange for verification.

The information requested will depend on where the customer lives and the laws there.

Once this KYC documentation has been verified, users will be prompted to connect a payment option.

At this point, the user can connect their bank account directly, or use a debit card or credit card. There are varying fees for deposits via a bank account, debit card or credit card and you should look at the pros and cons of each option. Generally, credit and debit cards incur higher fees, but may be faster in getting funds into your exchange account. Note that there is generally a limit (e.g., $2,500) for debit and credit card transactions, whereas wire transfers can be much larger.

3. Buy Crypto

Users can buy bitcoin or another listed cryptocurrency after choosing an exchange and connecting a payment option.

Cryptocurrency exchanges offer a number of order types and ways to invest. Most of them offer both market and limit orders.

- Market orders are transactions meant to execute as quickly as possible at the current market price.

- Limit orders set the maximum or minimum price at which you are willing to buy or sell, and then execute the trade at some future time.

For most people beginning their crypto journey, market orders may be preferrable.

Some exchanges also offer a way to set up automated investing — i.e., recurring purchases every day, week or month.

4. Store Your Bitcoins

Some exchanges have been "offline" during highly volatile events, which they have been heavily criticized for. Events may escalate panic, which can lead to lots of customers trying to withdraw funds at once — similar to how a bank run occurs. With crypto exchanges, these events can be triggered by things like lack of liquidity, excess lending and issues with collateral. Many individuals find it wise to be prepared in case a platform or company they're holding crypto with runs into these types of problems (or other ones).

Being true to the Bitcoin ethos would suggest that once a person has purchased some cryptocurrency, they should withdraw it to their own wallet. However, for beginner investors, leaving crypto in an account at the exchange is typically advisable until they fully understand the pros and cons of self-storage on a personal wallet.

For more on wallets and storage, check out *Part III* on Web 3.0. As always, do you your own research and reflect on your personal investing style, risk tolerance and values before making money moves.

* * *

BITCOIN SOUNDBITES

Now that we've explored the basics of Bitcoin, let's take a sweeping lens across some key crypto themes, highlighting content and soundbites from seasoned Bitcoiners and tech pioneers. These quotes are mostly from 2021–2022, and these speakers have remained notable over the years for their wisdom, assertions and bold moves within crypto and finance ecosystems. The one thing they all have in common? A passion for the value of the first viable cryptocurrency, Bitcoin.

JACK DORSEY:

"I didn't touch [Bitcoin] until 2008 when we started Square [Block]...but we encountered this crazy predatory system that was slow, that was obtuse...In 2009, you [could] see a chance to replace the whole foundation.

And everything that Elon was talking about in terms of ACH and the credit card networks...they have scaled, but they just are not relevant to today — and they're certainly not relevant to the future, especially when you consider the entire world and countries like Nigeria or Ghana or India — and its interconnection with countries like the United States and Canada and all over Europe.

What really drove my thinking and drives my passion around it is: if the internet gets a chance to get a native currency, what would that be in? And to me it's Bitcoin. Because of those principles, because of that creation story, because of its resilience...

What inspires me most is the community driving it, it reminds me of the early internet. It's the only reason I have a career."[9]

That makes two of us, Jack.

Background on the quote:

The "Bitcoin as a Tool for Economic Empowerment" live panel took place July 2021.

Known for his business acumen and crypto bullishness, Jack Dorsey was interviewed alongside Elon Musk and Cathie Wood by moderator Steve Lee on "The ₿ Word," published to YouTube by ArkInvest. He's the former CEO of Twitter, Inc., and he co-founded Block, Inc. Block is the developer of the Square payment platform.

While collectively their interest has been spread across various crypto projects, these influencers' principal support of Bitcoin has held strong. The discussion focuses on institutional adoption of the cryptocurrency.

Here, Jack Dorsey speaks to Bitcoin as a powerful network embraced by companies and nations around the world.

LYN ALDEN:

"I was cautious of Bitcoin. I basically explained in 2017 why I'm not buying it, and then I kept monitoring it — and then in 2020 when I bought it, that had some weight to it for people that were following my work because they know that I... saw information that changed my view on it. It wasn't bearish before but it was kind of neutral; I wasn't convicted enough...

Everyone should try to be objective at least; we all have human biases, but we should try to identify those."

—

"If you're on the Titanic and you think you're unsinkable, you're more likely to hit an iceberg... If you're on the Titanic, you still want to be looking out for icebergs."[10]

Background on the quote:

These quotes are from the YouTube interview "Bitcoin Risk

Assessment" with Lyn Alden, hosted by Peter McCormack.

Lyn Alden is an engineer, macroeconomist and financial writer who consistently delivers some of the strongest, most well-supported insights in the Bitcoin space. A quick web search for Alden or her company, Lyn Alden Investment Strategy, will lead you down several valuable crypto rabbit holes: insightful articles, a must-read newsletter, Lyn's poised social and on-stage presence, and her inspiring success story.

MICHAEL SAYLOR:

"I don't think there is any outcome other than favorable for Bitcoin."

—

"Everything that we're seeing in the last twelve months — inflation, war in Ukraine, Russian sanctions, the Canadian trucker crisis, CPI, unexpected PPI, and anxiety about how to manage a portfolio — all of these things have been bullish for Bitcoin.

...Although they're unfortunate and unpleasant for the world, they have underscored to every mainstream objective observer the use case for a global non-sovereign store-of-value crypto asset like Bitcoin."[11]

Background on the quote:

It's always nice to catch up on some of the yearly Bitcoin Miami conference headliners. Michael Saylor, co-founder and chairman of MicroStrategy, lets us know he's still bullish on the #1 digital currency during the annual Bitcoin conference in 2022.

This moment comes after the U.S. released an executive order on crypto, and Saylor comments on the anxiety around regulation and hopeful paths forward for Bitcoin. Interviewer Cathie Wood points out — crucially — some cautious notes on CBDCs (central bank digital currencies — centralized cryptos run by government banks). Wood and Saylor agree on the hopeful momentum they're seeing at this point in the bipartisan Bitcoin space.

* * *

PAUSE AND REFLECT

"There is no exquisite beauty...without some strangeness in the proportion." — Edgar Allan Poe

"How often have I said to you that when you have eliminated the impossible, whatever remains, however improbable, must be the truth?" — Sir Arthur Conan Doyle

Before we move on to explore other areas of the crypto landscape, we'd like to examine the road to Bitcoin adoption on a slightly more personal and academic level. While the financial aspects of Bitcoin provide a great deal of excitement, there's more to this journey than buying, selling and trading.

As we mentioned, Bitcoin is a network...and blockchain technology serves as the foundation for that network. We've touched on peer-to-peer payments, and how those may change the way individuals transact all around the world. But all of these use cases and functions of Bitcoin had to have started somewhere.

While learning about blockchain, we discovered that perhaps the most thrilling lessons were found under the hood, around the inner mechanics of cryptocurrency. The fascinating mysteries behind Bitcoin's creator, Satoshi, and this original crypto's early days remain unsolved. And, as many crypto supporters would suggest, that's for good reason.

It was necessary for this entity — which might have been one person or several — to stay concealed. And most argue that it is still just as important, and perhaps always will be...for safety, privacy and all the tenets that *cypherpunks* care most passionately about. That, plus maybe it's not at all about who created the technology. For as the infamous saying goes, "we are all Satoshi."

Cypherpunks are people who advocate privacy tech as a

means for social preservation and progress. Sound interesting? Let's explore the world of cyphers.

* * *

EDGAR ALLAN POE WAS BULLISH ON CRYPTO

My appreciation for Edgar Allan Poe began, like most, with "The Raven." I experienced the dim of the December room, the sounds and shadows, the hollow luxury of love beyond the grave. This was poetry. It was art with a maddening pulse.

Next came "Annabel Lee," just as somber, with a rush of tides and cold wind. Later I would read Poe's stories, which only pulled me further into his offbeat, gloomy chamber. I was transfixed by one particular copy of "The Black Cat"— letterpress printed with tattoo-style illustrations. "The Tell-Tale Heart" and "The Cask of Amontillado" rattled me to the core. (Spoiler: he definitely had a thing for insanity and people stuck in cramped and dangerous places. Cute.)

Of course, "The Masque of the Red Death" is perhaps the single tale with such meticulous foresight into our modern pandemic era. His writing remains a portal to a very secret, chilling space that some are only willing to witness for a moment...a place that Poe allowed to enshrine him completely.

So what does this all have to do with crypto? When I was first curious about the computer science behind digital currency, I took a quick tumble through Wikipedia to learn more about encryption.

As it turns out, Poe was more than the father of fright — he helped usher cryptography to the masses in the U.S. in the 1800s through stories, puzzles and prizes.

What Is Cryptography?

Cryptography is, at its heart, secure and private communication. In its earlier stages, it typically manifested itself as the art of encoding and decoding writings or ciphers. Broadly defined, it includes creating and analyzing protocols to avoid third-party observation — but in computer science, we focus on encrypting and decrypting electronic data, sent or stored. This involves math-based techniques like discrete probability and computational number theory.

In cryptocurrencies, it's used to allocate and utilize private keys (or addresses) within a network as transactions take place between individuals. This ensures the security and anonymity of senders and recipients and protection from double-spending — a primary pitfall that any viable digital asset must avoid. Bitcoin was one of the first digital currencies to use cryptography and do so successfully.

In cryptography, we typically illustrate transactions using "Alice" and "Bob" to represent two people sending and receiving encrypted data. With secure cryptography, these two identities are allowed to remain anonymous and go about their business without a third party (company, surveillance, hacker, you name it) watching or getting in the way.

But as I mentioned, cryptography was used prior to computer systems in sending and receiving secret messages. It's a practice that's been around for thousands of years. More recent examples include Renaissance Vigenère ciphers and rotor machines in WWII.

Contests and Challenges: Solving Cryptograms with Poe

So now that we've laid all that to rest, let's revisit Poe.

In 1839, Poe was working for the magazine Alexander's Weekly Messenger when he received a riddle sent in by a writer. Naturally, he found a way to make it better — and included several additional riddles alongside the reply. His improved method for encrypting puzzles included a definition of what today we call substitution ciphers.

For the next several issues, Poe challenged a larger audience of cipher submissions, saying he could solve anything that came his way. Poe corresponded in earnest, solving puzzles that stumped the masses while publishing swift replies to each challenge.

Most notably, he outlined criteria of what makes for a true single substitution cipher versus clues that indicate that they're fakes — and showed why. All of this was a big deal for the time, back before Alice and Bob.

Next, he published in four installments "A Few Words on Secret Writing" — a rare and clever roadmap for encrypted text. A Satoshi-like move, some may say. The final cryptograms featured in this series, "The Tyler Cryptograms," remained unsolved for over 150 years.

In 1843, Poe entered a short story called "The Gold Bug" into a writing contest. (I find it comforting to know that even the king of macabre had to submit his stuff.) Of course, he won. But what's most memorable about this is that he used a cryptogram in his piece that launched its characters on a treasure hunt. It made the story engaging (while adding his signature creep factor) — and was enough to tip the vote in his favor.

"The Gold Bug" also served as a key source of inspiration for William F. Friedman, who would become head cryptographer of the NSA.

These certainly weren't the only places where Poe included

secret puzzles. A posthumous poem of his, called "An Enigma," featured a hidden message encoded in the piece. And while many more authors of his time and thereafter included hidden messages in their works, Poe's bullishness on cryptograms was arguably unparalleled.[12,13,14,15]

Encryption Today and The Bitcoin Community

If you're looking for computer classes to tickle your brain, or perhaps you're a crypto whiz who would love to get more technical with your expertise, I would definitely recommend learning the basics of cryptography. The class I'm taking is more challenging than I ever imagined and definitely a step up from my SQL introduction as an undergrad. But it is a lot like learning about secret codes and, therefore, extremely fulfilling — bigtime cypherpunk vibes. It's also just helpful to have a bird's eye view (quoth the Raven) into the structure of digital currencies on a fundamental level.

I hope these topics don't feel like too much of a stretch, but I really believe that some of blockchain and crypto's greatest gems lie at the crossroads of literature and technology. What is the future without the past? What are zeros and ones without art?

Poe's life was a testament to ingenuity and intrigue, and despite his struggles, he left an incredible mark on the world.

His genius explores dark territories of the mind and of society where most people don't dare explore.

It's an attitude that most Bitcoiners I know share: the bravery to explore the unknown and adventure to great heights, perhaps risking immediate security or comfort along the way. The crypto community is tied together by belief in a better charter for humankind — one that embraces curious ideas which only a brave imagination can afford.

Plus, we often work in very cramped quarters.

I'll leave you with a thought. Is it possible that Poe's final substitution ciphers — solved by Terence Whalen, Jim Moore and Gil Broza — could have any connection to the inception of Bitcoin? I have a couple of wild theories...what do you think?

PART II

Altcoins: You Win Some, You Lose Some

PAUSE AND REFLECT

The term "altcoin" refers to any cryptocurrency other than Bitcoin.

While Bitcoin — as a tech protocol and form of currency — remains our primary focus, we feel it's necessary to have a sweeping view of this entire landscape in order to assess each crypto for its merits, value or flaws.

Our aim is to provide you with information to make your

own decisions and learn about the digital asset ecosystem holistically. Ultimately, the backbone of our education (and, we believe, any good one) lies in deep knowledge and understanding of Bitcoin as the original standard against which all cryptos may be measured.

Bitcoin operates in the crypto space as a type of money that, as we've mentioned, presents value as a potential reserve currency for the entire world. It's pretty special. Altcoins are tech innovations that serve a variety of different purposes, and many of them aren't created to become a widely accepted currency in this way. They can be good for lots of other things, though, and the following chapters will explore a small piece of that.

We like to see Bitcoin and some altcoins as complementary... or, at the very least, they occupy an overlapping space while bringing unique attributes to the table.

If you're looking to delve deeper into the nuanced properties of Bitcoin and understand how other blockchains compare, we recommend reading *Bitcoin vs. Altcoins* by Phil Champagne. There are a number of fantastic reads on Bitcoin more generally, and we'll summarize a handful of them in an upcoming section.

* * *

THE ALLURE OF ETHEREUM:
SUPER PLATFORM FOR TECH INNOVATION

We hope you found value in our introduction to Bitcoin and its utility. There's plenty more to talk about regarding Bitcoin…but it's important to know that everything discussed so far also lays a foundation to better understand the second largest crypto asset, one that has grown in popularity since its public launch in 2015: Ethereum.

This piece will kick off a series on Ethereum and its related services and applications. It's a tall order, but I'll do my best.

While Bitcoin is championed by many enthusiasts (often called Bitcoin maximalists, or maxis) as the one truly decentralized cryptocurrency, it's still not the only crypto in town. In order to grasp the wider landscape of this digital ecosystem, familiarity with Ethereum is key. If you've wondered what NFTs are all about or what altcoins like Aave or Polygon bring to the table, this is the place to start.

Ethereum's Early Days and What Makes It Unique

Our story begins with Ethereum's co-founder, Vitalik Buterin — a Canadian-Russian programmer and all-around genius. He wrote for several publications in the crypto space and co-created Bitcoin Magazine in 2011. Just a few years later, he published a white paper that outlined the importance of a distributed software platform built on blockchain technology.

The Ethereum developers set out to achieve this by offering, in Buterin's words, "a universal programmable blockchain and packaging it up into a client that anyone can use." This "world computer"[16] would follow in the footsteps of BitTorrent and Bitcoin by establishing a digital platform built on transparency and trust.

Using the coding language Solidity, it would serve as a hub of growth for decentralized applications (dApps) by extracting points of failure and restricting censorship, as well as offering a haven for peer-to-peer messaging, browsing and networking. The long-term picture? Distributed governance and enhanced technological collaboration — sparking unlimited potential.

How Bitcoin and Ethereum Differ

There are similarities here to Bitcoin; namely, they're both networks built on top of blockchain technology. In both networks, miners are rewarded for transactions that are cryptographically signed, but Ethereum miners are paid in tokens called ether (ETH). So while Ethereum itself is a platform, ether is the currency that provides the incentive to keep it running efficiently.

There are, however, many differences. First, we've learned that Bitcoin as a currency is scarce because its supply is capped at 21 million. Ether, on the other hand, has no set limit on its supply. The monetary implications of both networks are thus notably distinct.

The most important technical distinction comes down to a consensus mechanism called proof-of-stake. "Consensus mechanism," as we mentioned earlier, is a fancy way to describe what the computers in a network must agree on in order to work together securely. In contrast to Bitcoin's proof-of-work consensus mechanism (in which miners use energy in order to produce new blocks), Ethereum secures its network by following an algorithm that allows validators to vote for the outcome.

To explain this further, let's use a word analogy formula, like the ones you might find on a standardized test: "validators" are to Ethereum what "miners" are to Bitcoin.

Validators : Ethereum Blockchain :: Miners : Bitcoin Blockchain

Now, think of the word "staking" to mean locking up some cryptocurrency for a certain period of time. With staking, the ether (ETH) token serves as a form of collateral. Validators stake capital into a smart contract that runs on the network. (We'll explain smart contracts in the next chapter — don't worry.)

The idea is that if validators lock up their tokens in the network, they'll receive a reward for doing so.

Many argue that this move to proof-of-stake consensus has made the Ethereum network *more centralized* due to a skewed distribution of ether (with top addresses holding a significant percentage of tokens), as validators are ultimately managed by a small number of actors.

Buterin knew he couldn't create a proof-of-stake network right from the start — he had to begin with the proof-of-work foundation and build from there. In an event called "the merge," which occurred in September 2022, Ethereum began its gradual shift to the proof-of-stake model.

The transactions on the Ethereum network also cost fees, commonly referred to as gas. Gas fees are essentially the cost to compute or process a validator's transaction. The account sending the transaction will be debited this amount — and (metaphorically) like the fuel we use to fill our cars, the price of gas can fluctuate. In the case of Ethereum, gas fees can rise and fall a significant amount within a given day.

Network Structures and Forking

As for structural differences: where Bitcoin operates as one distributed, trustless transaction history, Ethereum is more like a single public ledger technology. It's often referred to as a machine or supercomputer that can be used to create new programs.

Ethereum runs on thousands of computers spread globally, known as nodes — but it has fewer nodes than Bitcoin (aka fewer systems of computers in the network). In Ethereum's case, the software that allows nodes to read blocks on the blockchain is called a "client." Trustless connections give both networks an extent of decentralization…but where each lies on that spectrum is a separate topic that we'll save for an upcoming chapter, "The Newbies' Guide to Web 3.0."

One key attribute that's the same for Bitcoin and Ethereum, beyond their blockchain foundation, is that both protocols can technically be altered. When there's a disagreement on how to proceed given a certain development error or challenge, teams of developers may split into two different networks. If they decide to upgrade or make a change to the code, there will be what's called a "fork". This has already happened in 2016, in the case of Ethereum Classic.

Up Next: Smart Contracts, DeFi and dApps

Ethereum was designed to allow decentralized applications to be built on top of its infrastructure. This is perhaps the most innovative aspect of Ethereum. It's a big deal because it provides a space for so many new types of interactions. Current use cases for dApps are decentralized finance (DeFi) and NFTs.

Programmable smart contracts are what make much of this innovation viable in the first place, so let's explore these a bit before diving into all of the exciting work being built on Ethereum.

* * *

SMART CONTRACTS: IF X, THEN Y

The term *smart contract* often gets tossed around when talking about the Ethereum blockchain and DeFi. This makes sense, since Ethereum is one of the most popular cryptocurrencies implementing this type of technology. Let's break down what smart contracts are, how they operate and how they contribute to the world of digital finance.

If you remember, we defined the Bitcoin network as operating on a cryptographic protocol. Smart contracts are a type of protocol that act upon certain agreed terms. They're conditional — just like the statement, "if X, then Y." Input leads to output.

In this way, smart contracts are pieces of computer code programmed to produce a predetermined outcome. They allow us to automate events, actions and results. It's important to note that these actions cannot be reversed or deleted — they're immutable.

More simply, smart contracts can be defined as a trustless agreement or promise made digital.

Permissionless and Full of Potential

These contracts run on the blockchain, and a crucial component of this technology is that they can be created without a third party — so no central authority needs to approve of the transaction.

Traditionally, banks and financial exchanges might authorize or mediate transactions, but with smart contracts those institutions can be taken out of the equation. With the help of smart contracts, developers can build decentralized applications (dApps) that run peer-to-peer.

At their core, dApps are software programs that run on a decentralized network of computers. This makes them typically inexpensive to operate and scalable, two ideal scenarios for growth and utility. The permissionless nature of dApps allows them to operate outside of anyone's control, meaning they're more resistant to censorship or restrictions.

It may sound pretty straightforward, and it sort of is...until we open up all of the possible use cases, caveats and consequences of said contracts. We then see how incredibly useful they can be — and also what would need to be done to ensure their effectiveness, safety and right application.

Where and How Smart Contracts Can Be Used

Many smart contracts run on the Ethereum network, as mentioned earlier, using the programming language Solidity. These if/then programs allow for "promises" to be executed on the blockchain — which could allow for exchanges of value and property without a third party. Since many nodes are witness to this transfer, both sides of the exchange could expect a seamless, and often faster, outcome.

Uses cases for Ethereum-run smart contracts include flash loans (where an asset is borrowed and paid back in a single exchange), token swapping (trading between currencies on decentralized platforms) and certain downloading permissions.

Other popular or potential areas where smart contracts may be used include real estate, financial services, the Internet of Things (IoT) and non-fungible tokens (NFTs). We have a chapter all about NFTs coming up ("NFTs in a Nutshell"), so it may help to remember that smart contracts play a role in these much-loved digital assets.

Legal Applications, Supply Chains and Oracles

Let's take a look at one potential use case outside of crypto:

wills. These legal statements outline what assets a person leaves behind, and because they have lawful implications, they're subject to third-party decision-making and judgment — and sometimes even mistakes or crimes. If they operated on smart contracts, several circles of people could be removed from the process as funds or assets would be transferred more automatically to the intended recipients.

You may be wondering how this entire process would function with as few people involved as possible. Likely, it would take a little while to get there. We've been accustomed to structure in our legal process that operates around a human-built system of ethics (i.e., law and order).

While it's an ongoing advancement, there are newer tools and developments beyond this process that may bridge the daunting gap between people and smart contract protocols. Take oracles, for example, which intercept data for smart contracts from the outside. If the source is deemed as trusted, they permit this data to be fed to the smart contract.

There is some debate over whether oracles are the most trust-worthy solution — and how they should be best used. Food supply chains are one example of this conversation in play. Wouldn't we need a human factor to check the product at certain intervals in order to ensure safety of all types of food before they reach the dinner table? Many argue that some combination of oracles and human actors would be needed to complement or truly fulfill the work of smart contracts.

* * *

THE DEFI DEEP DIVE

So far in our altcoin section we've talked about the origins of the Ethereum network, as well as its purpose and overall value in the crypto ecosystem. This platform allows for decentralized applications to be built on top of the blockchain — extending the possibilities of ownership, creativity and finance in the digital world further than ever before.

We've also discussed smart contracts, promises forged in immutable code that allow for agreements to be made outside the influence of humans (on Ethereum or other blockchains). These concepts will all form a bridge to the next topic at hand: how DeFi operates and what that means for society.

What Makes DeFi Different from Traditional Finance?

Decentralized finance, in its simplest form, refers to financial behaviors and structures that are situated outside of centralized institutions (like banks, governments and people in general). In the systems most people are currently used to, banks are not just a place to put your money; they're the custodians of it, which means they can technically limit your abilities, place costs on certain fiscal actions and even manipulate monetary behavior. Yet most of us have grown so used to these traditional methods that we couldn't picture it operating any differently.

In tech and crypto spheres, there's a phrase often tossed around that goes something like, "code is law." The impact and responsibility of software is now so profound that it can shape human capability and restraint, and this slogan serves as a reminder that decentralized technology uses protocols to stand in for the institutions themselves.

So those smart contracts we spoke about earlier — they're part of the package that can allow financial transactions and decisions to operate outside of banks and the Fed. Together, blockchain technology, cryptography and smart contracts form the foundation for DeFi.

Arguments for DeFi versus traditional, centralized finance? It has levels of censorship resistance. It's cheaper and often faster. DeFi can also be made more accessible to the general public, and no one bank or business sets the rules (note: this one's a slippery slope, since dApps are often built by people — and proof-of-stake could shape a less-than-equal playing field).

DeFi Put into Practice

So what are some examples of financial practices and behaviors that may operate differently in a decentralized setting?

Trading and Exchanges

- We've heard of centralized exchanges like Gemini, Coinbase and Kraken. But there are also decentralized ones (referred to as a DEX), where individuals can transact without the need for a bank or institution. The use of immutable code is more prominent on DeFi platforms, and that code is accessible to everyone. These exchanges — like Uniswap and SushiSwap — also offer a larger selection of tokens than centralized platforms do.

Stablecoins

- Stablecoins are a type of cryptocurrency tied to an external reference (like fiat currency, another cryptocurrency or an exchanged commodity). The value is then backed by the asset to which it's matched. As demand increases, new stablecoins can be minted to stabilize the price.

An example of this would be Tether, a coin pegged to the U.S. dollar (each worth 1.00 USD). Stablecoins are supposed to provide a reliable way to trade outside of a centralized exchange. However, having many unregulated stablecoins in circulation could make it difficult to discern whether they're truly backed when they say they are.

Lending and Borrowing

- When we lend or borrow money, institutions and legal systems are all part of the process. Banks and investors can be as selective as they want with their decisions. With smart contracts, these processes can be automated.

Aave and Compound are two examples of open-source and non-custodial protocols that run on Ethereum. They allow users to earn interest on deposits and pay interest on borrowed funds, with no human intermediary.

Loans can be collateralized against stablecoins or other cryptocurrencies whose prices fluctuate. Users should be careful as to which exchanges they choose to send their crypto, because with high reward comes high risk.

Financial Products

- With DeFi, traditional financial products can be remixed to the tune of a smart contract, removing large institutions and operating on fixed protocols instead. One example here would be insurance. Smart contracts could be used to substitute code for the insurance company itself, allowing all parties to insure their assets with fewer people involved, and potentially save some time. Funds are allocated exactly as specified in the contract. As discussed in an earlier newsletter, there are lots of factors that can be considered (like oracles) to ensure accuracy.

The Wrap-Up

From flash loans to margin trading, this list only skims the surface of DeFi possibilities...but we'll pause here for now.

The takeaway is clear: DeFi continues to make waves in both crypto and traditional sectors. As blockchain technology and decentralized innovation create unique financial opportunities, digitalization will continue to change how we all do business. With the help of smart contracts and the Ethereum network, DeFi is well on its way.

Still, it's important to consider the steps society will need to take in order to bridge current habits and customs with this fast-paced tapestry of tech — and the choices we each can make on a personal level to bring ourselves closer to financial freedom. DeFi demands a high tolerance for risk and the ability to be comfortable with fewer guardrails, so it may not be for everyone. Even so, the innovations in this space are showing us something: in the world of finance, there's room for a little disruption.

* * *

BEYOND BITCOIN:
TOKENS THAT JOINED THE PARTY

Whether you're new to crypto or not, there's something newsworthy burgeoning in the space every day. And while things felt a lot simpler when Bitcoin was the only prominent digital currency, a lot has changed over the years. Not only is blockchain being used for all sorts of exciting projects, lots of them have their own tokens — making them part of the larger crypto ecosystem. For some, it's a reality they'd rather not bother with. For others, it's a thrill.

One thing's for sure — if Bitcoin set the stage, altcoins came to dance.

It would take too long to touch on all the top assets here, but I'd love to give a little background on a few hot headliners from 2020–2022. I hope to speak to the value of specific projects and how they might contribute to this ecosystem — not as blind investments or short-term pump-and-dumps, but simply as tech innovations that carry novelty and potential.

Below I've listed a few projects that each play a distinct role in tech, DeFi or gaming. My goal is not to endorse or disparage any of them, just to shed some light on the crypto market beyond Bitcoin. None of this is financial advice.

There are thousands of tokens out there, so keep in mind this is a small slice based on things like public interest level at the time of writing, unique use cases, and overall "buzz." If you're looking to invest in anything, always do your own research first.

The Brave Browser and Basic Attention Token (BAT)

The Brave browser, developed by JavaScript creator and Mozilla co-founder Brendan Eich, integrates blockchains to provide a fast and secure web browser that blocks tracking software and unwanted ads. Personally, I enjoy the overall browsing experience and the feeling of privacy it gives me (i.e., notifications when sites want certain info). Users can also earn BAT tokens from watching ads — think Web 3.0 and the future of DeFi. It's all important to consider as we restructure how our attention and data are used.

Some may caution that Brave's security capabilities do not yet measure up to those of more mainstream browsers like

Chrome. But advocates of the newer tech see promise; Brave makes it so that the ones learning are also the ones earning.

Cardano (ADA)

This open-source blockchain uses proof-of-stake consensus to confirm transactions. Its cryptocurrency token, ADA, is used to run the blockchain, exchange value on its network and potentially host decentralized applications (dApps). Cardano's staking mechanism allows users to hold these coins as collateral and delegate to staking pools to earn potential ADA rewards.

NEAR Protocol (NEAR)

NEAR is an open-source platform on the blockchain with a built-in scaling solution. It aims to host dApps (decentralized apps) in a way that's fast and low-cost,[17] with community at its core. This proof-of-stake currency functions as a bridge to Ethereum, connecting builders to the greater collective. If successful, serverless apps and smart contracts could more easily connect to financial networks to store and transfer data for things like payments and lending services.

Solana (SOL)

Solana is a layer-one native chain cryptocurrency. Its use

cases exist in DeFi and gaming, and as a store of value. Solana is also fast; it rivals Visa in the amount of transactions it can complete per second (50k+).[18] This sacrifices some security but likely no more than many other similar blockchains. Solana is the first blockchain that claims to use proof-of-history,[19] a timing mechanism that allows for increased speed.

Worldwide Asset Exchange (WAX)

The metaverse is blossoming in popularity and sparking curiosity worldwide. WAX is a sidechain (think "sister chain" instead of a fork) of the EOS blockchain, and it was created for the sale and exchange of virtual assets like NFTs. It uses a delegated proof-of-stake consensus protocol — so token owners vote for delegates who make decisions. This can enhance scalability and offer profits through staking.

Popular games on WAX include play-to-earn gaming, such as the Uplift and Alien Worlds. For another major VR platform, you may want to research Decentraland (MANA) which has been growing in popularity for similar purposes.

Z Cash (ZEC)

Z Cash is a decentralized on-chain cryptocurrency (and fork of Bitcoin) that helps keep transactions anonymous by offering the option to shield transactional information.

This cryptocurrency uses two types of addresses: transparent (where the ledger history is public and viewable to all, as with Bitcoin) and shielded (which do not publicize past balances).

Z Cash uses the zero-knowledge proof, a system which maintains the privacy of the sender and receiver's past payment information when verifying data. This allows for the choice of privacy within the world of secure, decentralized transactions. Z Cash also has its own wallet. For more on traceability, it's worth looking into articles and newsletters that touch on this ongoing conversation.

The Wrap-Up

Of course, for some BTC maxis, this info may be irrelevant. If you're all set on novelty, there's nothing wrong with staying in the OG Bitcoin lane, paying no mind to the altcoins in the rearview. But a bit of educational framework can be valuable, if only for learning's sake.

One great resource for up-to-date news and relevant POVs on alternate crypto assets is Trading View — a platform with information on trading trends, market movement, and a sense of the general zeitgeist around certain projects. It's a space for chart analysis and learning, and it aggregates content on crypto tokens in a helpful and timely way.

We're beginning to see more innovation shaped around blockchain tech, dancing along the DeFi spectrum. And while Bitcoin remains top dog, I find it's worth it to keep tabs on the space holistically.

There are endless projects and topics out there to learn about — YouTube, CoinDesk and Substack are all great places to dig deeper into the value or drawbacks of alts. Whether you've been dabbling in investing or you just want to know what all the noise is about, there's always something new to discover.

* * *

IOT AND AN ALTCOIN CASE STUDY

In this chapter, we'll explain a new construct that goes hand-in-hand with crypto innovation: the Internet of Things. Our goal is to detail the purpose of IoT and what these inter-connected systems may offer to society as we continue to see growth in blockchain sectors.

Then, we'll look at one specific crypto project (Helium) and the scope of tech solutions it claims to provide. We will try to be as objective as possible, gathering some information from resources online. Keep in mind, though, that assessing new inventions requires a bit of digging (and a healthy dose of discernment), which is precisely why we want to conduct this experiment.

At the end of this chapter, we hope you'll have an idea of what the Internet of Things includes (though a few pages on one altcoin only scratches the surface of creativity in this space). You may also have a specific picture in your mind of what the Helium Network is and how it functions. Perhaps you'll get familiar with a few of the issues this project aims to solve.

Most likely, though, each reader will walk away with a unique take on whether this information feels useful or valuable. We believe that's a healthy approach — not only to this particular network, but to the world of crypto. Whether it's the leader of the pack (like Bitcoin) or a brand new coin you've never heard of, there's a saying in this community that we never tire of hearing: Do your own research.

What Is the Internet of Things?

In simple terms, IoT is a network that aims to improve the interconnectivity of physical devices and systems. And it could impact industrial and social infrastructure around the world in years to come.

Though it may sound futuristic, the Internet of Things is rapidly evolving and has emerged as a potential solution for exchanging data securely and efficiently via the internet.

An example of a familiar IoT product would be a smart device used in a person's home (like a security system) or a smart-watch that a person might wear. The commonality is that they usually incorporate a sensor or some type of software that allows communication with other objects, and this transfer of data allows the sending and receiving of information *without* extensive human intervention.

There are several categories of IoT systems. We won't give an exhaustive list, but to start, here are a few applications:

- Industrial IoT (IIoT) – e.g., agriculture and farm automation, manufacturing, energy systems)

- Commercial IoT – e.g., healthcare products, such as a blood pressure monitor

- Consumer IoT – e.g., smart home features, voice-activated appliances

- Infrastructure IoT – e.g., transportation and city management solutions

- Military (IoMT) – e.g., defense and surveillance tools, biometric tracking

Advancements in AI and cloud computing have given IoT industries a boost. As we see technology develop, we also notice greater opportunities for IoT product design and usage.

One argument for the utility of IoT lies in supply chain dynamics. Some argue that if we faced a large-scale supply chain crisis (which became more of a concern in some places following events like the COVID-19 pandemic), greater interconnectivity could target inefficiencies and speed up processes from production to storage and delivery.

IoT, they argue, may act as a high-tech conductor for this increased efficiency throughout a supply chain by collecting data, tracking items, identifying existing or lacking inventory, monitoring supply through enhanced visuals—and allowing systems to communicate each of these steps more transparently.

On one side, many call for greater privacy controls or standard-ization within IoT before we might see any promise. Others maintain that IoT can improve the quality of connectivity and the analysis of data, making it worth the effort and resources.

Helium's Potential Role in IoT

Helium is a network of wireless hotspots used to provide connectivity for IoT devices. Individuals are paid for setting up these hotspots in Helium Network Tokens (HNT).

While some coins try to capture and migrate legacy business, Helium taps into an emerging market: the Internet of Things. This industry includes objects or devices that are interconnected; they can communicate with one another (over secure connections) and use AI or automated services to track and use data — in turn, improving upon a process or product.

This network, dubbed the "people's network" by its supporters, serves to connect various devices that exist within the Internet of Things. Things like sensors and monitors, which each have an independent purpose (be it commercially driven, for industrial use or something else), may need to be aggregated so that companies or customers can access and organize all the data that's being collected (ideally in a secure way).

The Helium Network is made up of hotspots (like internet routers), and as we mentioned earlier, these locations serve as connectivity hubs for the devices that run on the Internet of Things.

The leverage of this "people's network" could be far-reaching. Having partnered with DISH,[20] Helium's hotspots have expanded into the 5G sector, and there are many who claim that it is the fastest-growing wireless network in the world.[21]

Vietnam's Nationwide Helium Network

Urban development tech company VIoT and IoT solution provider Kerlink seek to bring Helium's network to local communities in Vietnam.

This move may "make it possible for Vietnamese factories, businesses, cities and consumers to benefit from cost-effective, easy-to-deploy vertical wireless solutions..." says the CEO of VIoT Group. "...Think Airbnb or Grab for telcos: if I have an empty room, power and internet, why would I not use it as a network tower?" (Read or listen to more of this on Matt Turck's interview with Helium CEO Amir Haleem.)[22]

This could be used as a template or large-scale example of Helium's capabilities in other countries around the globe.

Solana and HNT

In the spring of 2023, the Helium Network migrated to the Solana blockchain. In other words, this project moved from its own blockchain to operate on another one in the hopes of scaling and transacting more quickly. The upgrade made the HNT token compatible with Solana's infrastructure and the platforms associated with it (again, to help with Helium scalability and utility).

This event was regarded by many in the community as a significant blockchain migration, with Helium users staking their tokens and voting on-chain. A majority of stakers were in favor of the shift, which allowed the migration to take place.

Helium Adoption and Expansion

With roughly the same energy output as a standard LED lightbulb, hotspots[23] allegedly offer opportunities to mine tokens at home without a huge burden on electric bills or the environment.

At the time of writing this original article, there exist over 700,000 Helium hotspots globally, with tens of thousands of new mining locations opening up each month. Some attribute this to the network effect: the more a product is used, the greater its value. Still, it's important to consider things like the supply and demand of a product to assess its potential value over time.

Of course, swings in prices of cryptocurrencies can be volatile, and all investment carries risk.

As you'll see, this particular project claims to offer industrial utility in the IoT space, but whether or not this sales pitch delivers on its potential remains to be seen. Like all cryptocurrencies, Helium is young. Some cryptos prove quickly to be scams, while others show promise at first, only to lose necessary support.

Several companies are taking a serious look at what the People's Network has to offer, and this recent burst of interest may clue us into greater potential down the road. DISH was reportedly the first major wireless carrier to deploy Helium miners as part of its 5G rollout — huge news in 2022 for the telco industry and HNT supporters. Leading asset management company Grayscale announced in March 2022 that they would consider including HNT as a product. Food and beverage company Nestlé has been cited as exploring Helium monitoring for its ReadyRefresh water cooler refill levels, while Victor® uses Helium tracking for mousetraps. There are potential applications across various industries, but the central use cases for Helium are device tracking and data gathering.[24]

As co-founder Amir Haleem states in an interview with FirstMark VC Matt Turck:

"Everyday things that we use shouldn't need cellular plans... the Helium Hotspot opens the door to an ecosystem of possibilities that allow people to connect anything from pet collars and ride-share scooters to sensors that monitor air and water quality."[25]

Who's Behind the Helium Network?

Another key question to ask when researching a crypto: Who's steering this project?

Helium's parent company is unicorn venture Nova Labs, previously known as Helium Inc. The name change comes with a March 2022 rebrand after the company raised $200M in capital.[26]

When we look deeper into the Internet of Things and the possibilities it opens in the world of tech, it's important to note that some big names in tech and investing are included in the Helium network's efforts. As we've learned in crypto, this does not always guarantee success (in fact, sometimes the opposite is true). Nonetheless, a top consideration in assessing a project's viability lies in understanding the motives and *motivators* behind its creation.

VC firms Andreessen Horowitz, Deutsche Telekom and Goodyear Ventures are a few of the varied investors backing the endeavor.

This crypto initiative is a collaboration between videogame industry exec Amir Haleem, Sproutling's Chris Bruce, and Shawn Fanning — co-creator of Napster.

Mining and Trading Considerations

Helium works off a proof-of-coverage model — where users can buy and set up their own hotspots to earn tokens and participate in the network.

While HNT may be bought or sold on exchanges, many investors who do their trading through leading apps may encounter a lack of access on their platforms of choice. Time will tell if increased adoption and interest will shift accessibility in these processes.

Drawing Conclusions

The creation of the Helium Network addresses unique use cases, separating itself from other altcoins by its project roadmap, global reach, familiar cast of characters *and* alignment to possible IoT solutions. The numbers are impressive. Yet many startups and projects carry immense early potential only to later fail. This can be chalked up to factors that aren't always readily apparent.

Plus, it's not always obvious which facts are true. Media coverage of crypto projects remains sparse given the nascency of this ecosystem, and even then, it can be difficult to discern advertising spin and carefully selected stats from cold, hard facts.

The framework of this altcoin allows us to take a closer look at some problems in need of solutions (think device communications in a world of rapidly evolving tech or even supply chain concerns), and that alone carries educational value.

I tend to think of alternative crypto assets (ones outside of Bitcoin) as each having their own unique use cases, capacities and even mini ecosystems. Some blockchain projects may flop, subject to feeble frameworks or scammy intentions, while others could potentially revolutionize our entire world. As builders, learners, investors or change-makers, it's up to us to decide not only whether a project has value, but whether we are motivated to support and sustain its cause.

IoT and the Growth of Tech

Whether you buy into the roadmap of the Helium Network or remain skeptical of its role, the birth of this crypto underscores why so many new coins are created in this space: blockchain technology could be poised to meet the needs of companies and individuals as we enter a new landscape of innovation.

The development of the Internet of Things boils down to the search for solutions to the *progress* that accompanies network effects. The growth of tech demands interconnectedness, and in turn, our interconnectedness demands new tech. We are seeing a cyclical reaction in real time to human adaptation... and while not all altcoins will prove viable (or even honest), there could be a few that tout creative and industrial advantages. Just make sure to rely on thorough research.

* * *

TERRA'S TUMBLE

Inflation's still soaring. Terra may have met its maker. And Bitcoin's hanging in there like a prize fighter that's seen this kind of blow before.

Man, you miss a week of this stuff and it's like you can never catch up.

Let's start with Terra, the coin that plummeted nearly 100% almost overnight. I'm not going to get into the weeds here, but a quick definition of what the asset is and why it's been controversial may be helpful for some.

What are Stablecoins?

Terra (UST) is an algorithmic stablecoin. Let's break this down. A stablecoin is a cryptocurrency that is pegged to something outside itself — be it another crypto or fiat or a commodity — and this external asset is what backs its *value*.

Now for the algorithmic part: these coins were intended as a solution for keeping the token value stable by increasing the supply when market value goes up, and lowering it when down. As demand increases, new stablecoins can be minted by burning a partner token, Luna, and sending a small portion to the treasury.

Take note of this attribute because it plays a key role in some of the recent controversy, and it's a big part of what makes stablecoins important.

So why would people support stablecoins? Well, they are trade routes that can be used if someone doesn't want to rely on an exchange — and they've provided opportunities for staking and earning yield, outside traditional institutions or banks. (See our DeFi chapter for more on all this.)

Because they're *meant* to be stable, stablecoins often operate as havens for funds when trading in a market that's used to seeing ups and downs.

They may serve as a fundamental bridge between cryptos, and people have been hopeful that this algorithmic mechanism could enable a future where even stablecoins could be decentralized. Basically, they could be helpful in a lot of ways. In theory.

What Happened with UST?

Terra grew fast and feisty, reaching a $10B market cap by the end of 2021. (Market cap, or market capitalization, is the value of an asset multiplied by the amount of it in circulation.)

A panic was triggered, causing the Terra network to collapse. The stablecoin UST lost its peg to the U.S. dollar, causing a rush on the ecosystem. On the heels of the UST crash, crypto as a whole saw a $1 trillion loss in value.[27]

Many are trying to speak out to let people know that not all stablecoins behave similarly, but Terra's swift plunge has left many others wary.

As Lyn Alden explained in one of her tweets, this whole ordeal kept getting worse because of how it was managed during the crash:

"Terra's multi-billion dollar algorithmic stablecoin UST blew up today. Aside from destroying the value of $LUNA, they used their bitcoin reserves to try to defend the peg, kind of like a flailing emerging market using its gold reserves to defend its FX." —@LynAldenContact, May 2022.

Back to Bitcoin

Though it may be a surprise to some, many Bitcoiners don't seem riddled with anxiety about the late 2021 BTC price drop, despite the headlines at the time. This is typically because we see Bitcoin as a long-term store of value. This *lack* of anxiety comes down to things like time preference, risk tolerance and a person's overall attitude toward crypto ownership or investments.

Yes, institutional adoption and large-scale acceptance of Bitcoin are crucially important in seeing healthy monetary communion worldwide. But there's a tremendous amount of promise in its intrinsic worth — those attributes that make Bitcoin so incredibly different from other forms of currency or other investments. Remember: it's a network and a technology.

If enough people find value in the benefits of this digital asset, its network effects could allow Bitcoin to serve a unique purpose in the global financial landscape. It may transcend many of the pitfalls altcoins have fallen prey to — proving itself as a long-term store of value and widely adopted global currency. Only time will tell.

In Michael Saylor's words,

"Bitcoin is not just an asset, it's a network and it's a protocol... it's decentralized, permissionless, global, immutable, scarce, auditable, instantly transferable, not seasonable, highly divisible...mostly everything that gold can't do."

—Michael Saylor, Stansberry Research, "Bitcoin vs Gold: The Great Debate with Michael Saylor and Frank Giustra," April 22, 2021. YouTube.

PART III

Web 3.0: Decentralizing the World Wide Web

THE NEWBIES' GUIDE TO WEB 3.0

What Is Web 3.0?

Web 3.0 is an emerging concept that encompasses ideas and practices, related to a decentralized internet, that are now coming to the fore — and will be increasingly important in the years to come.

It's important to understand what we know now, so we can take advantage of Web 3.0 as it grows in significance. Let's start with a definition of Web 3.0 as it exists today and dig into some of the ideas behind it to examine why it matters so much.

Web 3.0 represents the next stage of the internet, where decentralization, digitally native money, autonomous systems and self-sovereign identity combine to enable new economic models operated by communities rather than corporations.

The Journey from Web 1.0 to Web 3.0

One good way to understand Web 3.0 is to compare it with what came before during the Web 1.0 and Web 2.0 eras.

> 1. The Web 1.0 era was populated by static news sites, personal websites and blogs. Users consumed content and data from these sites, but they didn't do much else. Communication channels were usually one-to-one interactions, such as email, and few monetization models existed.

> 2. The Web 2.0 era was characterized by a shift to user-generated content that could be shared via popular social media sites, marketplaces and web apps. In this way, communication turned from one-to-one to one-to-many. However, these interactions usually went through centralized entities, such as Facebook, Amazon or Google,

which were able to build lucrative business models around this activity.

3. Web 3.0 has the connectivity of Web 2.0, but it is achieved in a truly peer-to-peer manner, with no need to go through centralized entities. Instead, decentralized networks built around community-driven incentives enable users to interact directly and to benefit from the economic activity that takes place.

So...What's the Point of a Decentralized Web?

The idea here is that emerging tech (be it smart contracts, dApps or something that hasn't been invented yet) might allow for new types of connection *without* relying on a nucleus of control. But the question remains: exactly how decentralized are the tech systems underpinning this vision?

We'll share our thoughts on this in just a second, but let's first explore the possible advantages of a new, improved cyberspace.

There are a number of benefits associated with the emerging Web 3.0 paradigm:

User-Centered Economic Models

While the Web 2.0 era was characterized by a small number

of tech giants profiting from the activity of billions of internet users, Web 3.0 enables users to gather in globally distributed networks and generate income directly by providing services the community needs.

Personal Data Protection

The Web 3.0 model vouches for individual control of data in our ever-changing world of tech. This goal discourages entities from profiting from the exchange of personal information in the way Web 2.0 tech giants have. Users can operate anonymously or pseudonymously, with zero-knowledge proofs used to verify information without it needing to be exchanged.

No Single Point of Failure

A main reason that Web 3.0 is built on decentralized networks is so that they might be resistant to attack and, ideally, impossible to shut down because there is no single point of failure. This means if one part of the network fails, the other parts continue as normal.

Open to Anyone, Anywhere

Open-source software allows participants to view, access, distribute or use data regardless of who they are or what they own. It's the ultimate collaboration tool in the world of tech.

We believe the best example of open-source software created to date is Bitcoin. Bitcoin serves as a model for incentivization of business and commerce in a world not run by fiat currency. Its decentralized, public network underpins the potential for technology in years to come.

While Web 3.0 encompasses a much broader web of block-chains and aspiring tech, the goal and spirit of this community echo a similar ethos: mutual wellbeing does not require permissioned access through a central entity.

We've seen Bitcoin open up economic activity in developing parts of the world, and even help dissidents to continue operating within authoritarian regimes. And we believe this framework would allow a fuller picture of Web 3.0 ideals to reach their true capacity.

Of course, these plans are full of promise and creative use cases for crypto. They integrate a collection of technologies (like the smart contracts we've mentioned in previous chapters, or brand-new alts and DAOs) that bring the idea of permissionless, open-source tech to light.

While some argue that in a Bitcoin world, there might not be room for all of this tech to thrive, we'll have to wait and see what happens. The cool part is in knowing that any of this innovation is possible.

The Role of Blockchain in Web 3.0

Bitcoin is a decentralized blockchain network that was purposefully built so it cannot be shut down.

A blockchain is a way of storing information on a decentralized and distributed ledger, while blockchain networks are lots of individual computers coordinating to keep this ledger up to date and secure from attack.

Each blockchain serves a unique purpose, and therefore may add to the Web 3.0 ecosystem in different ways.

The Web 2.0 that we see today is a network of networks, but most of these networks are controlled by centralized parties. Just think of Amazon's, Google's and Microsoft's cloud systems and the vast array of popular applications they support.

Web 3.0 is different because it involves blockchain networks (rather than private corporations) interacting with each other through a process known in crypto circles as interoperability.

Remember: not all blockchains are decentralized, and each serves a unique purpose. Many of these networks may *complement* one another (or complement Bitcoin specifically), and that's where we begin to see potential across a connected, interoperable landscape.

The Architecture of Web 3.0

The role of blockchain in Web 3.0 changes how applications are built, with the Web 3.0 versions known as decentralized applications or dApps.

The basic architecture of a Web 2.0 application includes the browser that the user interacts with, front-end code for displaying data in the application, back-end code for retrieving that data and a database for storing it.

The first part of this is the same in Web 3.0, as the user interacts with the browser, and the data is displayed using front-end code. However, one of the main differences is the use of smart contracts. These programs exist on a blockchain and execute automatically if certain parameters are met.

Web 3.0 Business Models

As with everything else in crypto, new business models are emerging as we learn about the web 3.0 space and its infrastructure. Parallels can be drawn to the early days of the internet, where individuals and companies adopted the tech at the moment they adapted to it. However, some notable examples already exist.

Token Ownership

The most obvious economic difference between Web 3.0 technology and Web 2.0 technology is the accrual of value into tokens rather than equity. In Web 3.0, decentralized blockchain networks have a native token or coin rather than shares.

Fees on Use

A valuable token will have some utility in the network, such as paying validators for their work in confirming transactions. Tokens may also have governance rights attached to them, allowing the holder to vote for or against protocol changes. Finally, token holders could receive fees related to the protocol's main activity, such as lending, trading/exchanging or yield generation.

DAOs

Decentralized autonomous organizations or DAOs are a Web 3.0 amalgamation of communities, corporations and cooperatives. Like decentralized applications, they exist on blockchain networks and are coordinated via smart contracts. DAOs are usually established to achieve an objective, such as funding projects or investing to achieve returns.

Where Does Bitcoin Fit within Web 3.0?

"You don't own 'web3,'

The VCs and their LPs do. It will never escape their incentives. It's ultimately a centralized entity with a different label.

Know what you're getting into..."

—Jack Dorsey (@Jack), December 2021, Twitter (X).

Web 3.0 is the first internet era where truly digitally native currency is possible, and Bitcoin is the preeminent example of this.

In fact, some of the most prominent corporate supporters of crypto have put their full weight behind Bitcoin. This includes Square (now called Block), the digital payments company, which has established a cryptocurrency and decentralized finance business unit.

While a number of these innovators deny the legitimacy of Web 3.0, they recognize the promise of Bitcoin when it comes to payments and financial technology. For this reason alone, blockchain technology has proven its worth.

Other Bitcoin advocates believe Web 3.0 *has* to be built on Bitcoin because it is the most decentralized crypto network. For them, decentralization either exists or it does not, while others argue decentralization is a spectrum on which many networks can exist.

We will have to wait and see, but a future that includes multiple interoperable blockchain networks seems likely right now. There's nothing else that can do what Bitcoin does, and at the same time, the world of blockchain appears best poised to create opportunities in support of this cause.

* * *

VALUE ACCRUAL AND BUSINESS MODELS IN WEB 3.0

Having introduced the big ideas behind Web 3.0, I want to dig deeper into the business models that are emerging within this new paradigm, so women are able to take advantage. Before I do though, it's worth reminding ourselves of Web 3.0's fundamental aims and how these improve the internet.

Web 3.0 aims to build on decentralized blockchain networks that do not contain a centralized entity who controls access. Therefore, there is no single party taking profits from the activity that occurs but instead a level playing field for all. In this way, Web 3.0 enables consumers and suppliers to establish direct connections and bypass the middlemen.

A Word on Tokenomics

Tokenomics, which is the study of token economics within block-chain networks, is a big subject — but it is helpful to understand the basics before investigating Web 3.0 business models.

Tokens were introduced into decentralized blockchain networks to incentivize behaviors that would benefit the network as a whole. In the case of the Bitcoin network, the 21 million bitcoins were to be used as a medium of exchange, with the mining process acting as an incentive to keep the network secure.

In Ethereum, the process of paying to execute smart contracts and confirm transactions is referred to as paying gas fees, which are denominated in the native token ether.

Business Models in Web 3.0

The main economic difference between Web 3.0 technology and Web 2.0 technology is the accrual of value into tokens rather than equity. In Web 3.0, decentralized blockchain networks have a native token rather than shares, with users accumulating tokens in a variety of ways.

Fees on Use

A valuable token will have some utility in the network, such as paying validators to confirm transactions.

Fees may also be levied in relation to activity that occurs on-chain, such as lending or exchanging of tokens. When this fee is awarded to token holders, it can be thought of like the dividend that equity holders receive from a company whose shares they own.

Fees may be distributed to liquidity providers (LPs), who provide their own tokens as collateral to a lending and exchange pool, or to the holders of the protocol's native token (or a combination of both).

Creating Digital Assets

The distributed ledger at the heart of a blockchain network records ownership of assets in a secure, immutable and auditable way.

This is one of the benefits that have made non-fungible tokens (NFTs) so popular with creatives today. These unique digital assets, which can be art, music, collectibles or other digital items, are created and recorded on the blockchain so the creator has full control of how they are sold and who receives payment.

In this way, Web 3.0 introduces an empowering business model for creators that stands in stark contrast to the copy-and-paste model for content that exists in Web 2.0

(where it's easier for anyone to lift content and profit from others' work).

Investing and Yield Generation

As interest in a decentralized blockchain network grows, the value of its native token can increase and so can its price. This is why many of the early Web 3.0 business models revolve around trying to pick the winners and invest in a native token early on.

The incentive design in blockchain networks means that Web 3.0 investing can also involve chasing these incentives to generate returns. This is known as yield farming, where investors scope the market for protocols offering rewards, which usually involve providing liquidity and receiving more tokens in return.

DAOs and Web 3.0 Governance

Decentralized autonomous organizations or DAOs are important for Web 3.0 because they are the mechanism by which individual entities coordinate within decentralized networks. In this way, they can be thought of as unique Web 3.0 amalgamations of communities, corporations and cooperatives.

Like decentralized applications, DAOs exist on blockchain networks and are coordinated via smart contracts. They are usually established to achieve an objective, such as funding a project or investing to achieve returns, but can also have a purely social or community-based agenda.

DAOs usually include some sort of online community within a popular messaging platform, such as Telegram or Discord. They also control a communal treasury, with a unique set of governance rules for spending the funds to meet the DAO's objectives.

As an online community that is focused on helping women understand decentralized finance, we're interested to see how DAOs develop within Web 3.0.

* * *

ADVANTAGES OF A DAO

It's the norm in crypto for certain topics to capture the whole community's attention at once and that is the case for DAOs right now. There's a lot of hype to sift through but there's also a huge amount of innovation going on as well.

In fact, DAOs could be the vehicle through which all capital and labor are coordinated in a Web 3.0 world. So, let's dig a little deeper into this important trend.

Remind Me: What Is a DAO?

As mentioned in the previous chapter, DAOs are a Web 3.0 amalgamation of communities, corporations and cooperatives.

More specifically, the acronym stands for *decentralized autonomous organization*, which, as it sounds, is a group that coordinates its activities in a peer-to-peer manner without any need for a central authority.

DAOs exist on blockchain networks and are coordinated via smart contracts. They are made up of crypto natives who have chosen to use the decentralized structure of a DAO, rather than a traditional structure like a limited liability company (LLC).

The Benefits of DAOs

Each DAO has its own goals, but there are some common benefits across all of them.

Global Collaboration

Just as blockchain networks allow anyone to participate in a new financial system that is open and permissionless, DAOs allow you to collaborate with anyone around the world without limits.

Capital Coordination

A DAO usually has a common treasury, which is spent to meet its aims and objectives. In this way, DAOs enable participants to pool their capital with others in order to have a greater impact.

Flat Structure

DAOs lack the hierarchical design of traditional organizations. Instead, DAOs usually have a flat structure, are coordinated using smart contracts and stick to the principle of "code is law."

The DAO Landscape

At present, some of the most common types of DAOs focus on investing, collectibles, culture and media. A few examples of DAOs in these categories are described below.

FWB

FWB or Friends With Benefits describes itself as "the ultimate cultural membership" and has been designed as a community in which Web 3.0 creators can come together.

SeedClub

SeedClub is an investment DAO that supports tokenized communities. It includes an accelerator and studio for launching and scaling promising social token projects.

JUMP

JUMP is a media-related DAO for those working in advertising, branding and the Web 3.0 creator economy. Benefits include meetups, events and access to Web 3.0 leaders.

How Do I Join a DAO?

The first thing to do is understand the DAO landscape and assess whether there's one that matches your goals and ambitions. If you find one, you have a couple of options.

If you're a developer, you'll be in high demand. There's a lot to do, from orchestrating smart contracts to meet the goals of the DAO to automating operational systems and processes.

If you don't have any coding skills, don't be put off. There's plenty of demand for writers, designers and creatives who can help to promote and grow these crypto-native communities.

* * *

DIGITAL WALLETS FOR CRYPTO AND WEB 3.0

Wallets are one of the most important parts of crypto.

They let you access, store and transfer all the digital assets you will encounter, from coins to NFTs. That's why we've written this short guide to help you make the most of them and to stay safe while doing so.

Whether you plan to buy some crypto assets, already own lots or you're dabbling...security should be a priority and never an afterthought.

What Is a Digital Wallet?

A digital wallet is a software system that allows you to store various types of digital assets. In the non-crypto world, common digital wallets that allow you to store fiat currencies in digital form include PayPal, Apple Pay and Google Pay. It may help to think of these as examples of how funds can be represented electronically in order to draw comparisons to crypto storage, as long as you keep in mind that bitcoin is fundamentally different from the dollar.

In crypto, wallets allow you to access your digital assets that are stored on the blockchain. Your crypto wallet lets you access or move the assets associated with your address, rather than storing the assets themselves.

Digital assets on a blockchain can be accessed and transferred via a process of public-key cryptography. This involves two keys, a public and a private one, and it is these keys that your wallet stores.

You could store your keys without using a digital wallet, by remembering them or writing them down.

Different Types of Digital Wallets

There are many wallets available and the market for them is ever-expanding. However, there are essentially two main types worth knowing about.

Hot Wallets

A hot wallet is one that is connected to the internet. They may exist as browser extensions, mobile apps or both. Some well-known examples include MetaMask, Coinbase Wallet and Trust Wallet.

Using a hot wallet has some risk attached to it because the wallet is always connected to the internet. This makes it more accessible to hackers, who may not be able to hack into the wallet itself but may be able to exploit vulnerabilities in your computer, operating system or browser to trick you into giving them access.

Cold Wallets

A cold wallet is one that remains offline. It is usually a small USB device, referred to as a hardware wallet. Popular examples include Ledger and Trezor, but there are others available.

The main reason why it's a good idea to use a hardware wallet is for increased security. The offline element means the number of attack vectors any hacker can use to steal your assets is greatly reduced. However, this increased security comes at the expense of convenience because there are more steps involved in transferring assets from cold storage.

Using a Digital Wallet for Yourself

It's common for people to keep the crypto they are *saving* in cold storage and the crypto they are *trading* in a hot wallet.

This allows them to try out DeFi, NFTs and Web 3.0 with their hot wallet, while keeping the majority of their investments in secure cold storage. Our caution is to start slowly and carefully, but don't be afraid to get to grips with how wallets work, as they're a fundamental part of the new crypto paradigm.

PART IV

NFTs and the Metaverse: Brave New You

NFTS IN A NUTSHELL

Our crypto adventure has taken us from the Bitcoin and Ethereum white papers to smart contracts and DeFi, and we've finally arrived at the world of .jpg primates that sell for $24 mill at Sotheby's.[28] This chapter is solely about NFTs, and we know you've been waiting patiently. So, without further ado, let's tackle some common questions that may be asked by the NFT newb or the avid collector who's just here for the fun of it.

Break it down for me! What is an NFT?

NFT stands for non-fungible token. Just like other tokens in the digital world, NFTs are assets that can be bought, sold, held and used. The non-fungible part refers to the fact that they are unique and not interchangeable with others of the same value. In this way, NFTs carry value in their uniqueness.

These assets live on the blockchain, and as we've discussed before, this technology records and stores information on a distributed ledger. In this case, the transaction of the NFT sale and ownership of said item are kept on the blockchain — establishing the person holding the NFT as the sole, true owner of it.

NFTs can appear in the form of images, videos, tickets, certificates, information and a whole host of other collectibles and goods.

When an NFT is created, it is minted. Minting is the process where the digital art is tokenized on the blockchain — ready to buy, sell, trade or enjoy.

Why would someone want to own one?

Think about the ways collectibles have operated in the tangible world over the years — maybe imagine someone you know who is a big collector of something. Pokémon cards?

Beanie Babies? Warhols?

Sometimes it's a small hobby, and other times it's a big part of their life or work. The items they collect may have a personal significance for the owner, or perhaps it's purely an aesthetic choice or flex. Often, these collectibles are worth money or have accrued value over time. Their value is derived from a combination of scarcity and demand. This works the same in the NFT world, too.

Perhaps the monetary value doesn't really matter as much to the owner, but the artist, creator or story behind that particular thing is where they derive meaning. The *community* surrounding it could present the greatest inherent value of all.

A person might collect something for any of these reasons — or all of them! It's similar with NFTs. People create, buy and sell NFTs for different combinations of those motives mentioned above, except the territory is so new and vast that a lot of these elements are amplified. The stakes may be a little higher and these collectibles are an entirely novel concept.

A key reason that people might care about NFTs so much is that they're changing the meaning of ownership to something entirely different — with an emphasis on creative sovereignty. Historically, many artistic endeavors have been known to represent a personal or even spiritual pursuit that doesn't always translate to dollars.

What's more, artists may feel their work is appropriated, misrepresented or misused (think of aaallllll of Taylor Swift's re-releases). Since the ownership of an NFT can't be mimicked, NFTs can provide a way for creators to truly own the content they've dreamt up...without a middleman or agent in charge of the marketing and bidding. This is especially the case when the platform where the art is produced or sold has some degree of decentralization.

But are they useful?

NFTs can go up (or down) in value. In this way, NFTs can be seen as investments. The total sales volume of the NFT market in the third quarter of 2021[29] was estimated at roughly $10.7 billion, according to DappRadar.[30]

Beyond monetary value, many people love to show off their NFTs in different ways or various arenas. They can serve as avatars, often in the form of a profile picture for social media and gaming. Twitter banners inspired by bold or heart-warming elements are yet another example of NFT reach and scope.

With all the buzz around metaverse realities, identity is certainly part of this wild ride...and the fact that people may come to identify with an NFT has so many implications (especially if they want it to represent them forever).

Play-to-earn games utilize NFTs by allowing users to interact with different types of NFTs in-game. You can have an NFT that acts as a "land key," giving you ownership or access to a digital property — or it can act as a virtual wearable item, like sneakers or jewelry.

From art and music to governance, gaming to virtual real estate, or tickets for sports games or concerts — the list of spaces for NFTs does not yet have an endpoint in sight.

Does owning an NFT mean I have the commercial rights to it?

The rights to distribute or commercialize an NFT are not necessarily included with the purchase of it. So if someone wanted to build a business using the NFTs they've bought, they would likely need specific licensing to do so. Intellectual property rights are occasionally part of the ownership package when a person buys an NFT, but often they're not — and this is important info to make sure of when buying them.

Are all NFTs on the Ethereum blockchain?

Nope. Although popular marketplace OpenSea carries NFTs that live on the Ethereum blockchain (remember those gas fees we mentioned? They're part of that purchasing process for Ethereum-backed NFTs, too), there are other blockchains like Wax (via atomichub.io) and Avalanche where a person can mint an NFT as well.

NFTs are also being explored on top of Bitcoin. This landscape is coming to fruition through exciting projects like DIBA (Digital Bitcoin Art), which builds on Bitcoin's layer-two protocols using RGB (Really Good for Bitcoin) smart contracts. You can flip back to the "What Makes a Network Good" chapter for a refresher on layer two and Lightning.

Bitcoin Ordinals are on-chain tokens inscribed onto satoshis. They can be thought of like digital artifacts, and similar to NFTs on other blockchains, they represent ownership (only in this case, they live on the Bitcoin blockchain itself.)

Interoperability is a term you may hear thrown around here, too, which basically means different blockchains working together. As we look to the future, interoperability could become more desirable in the world of creating, selling and owning NFTs.

Where are NFTs bought and sold?

Popular marketplaces include OpenSea, Rarible, AtomicHub and more. There are smaller, more curated spaces like Nifty Gateway, MakersPlace and SuperRare. Then you have brand-led experiences (like NBA Top Shot) that function as creator, inspiration and marketplace all in one.

A lot of NFT action happens in-game, meaning people can buy or trade NFTs within the world of a game while they're playing it. Experiences and items can be tokenized to become part of that environment, and this is something we're seeing more as the metaverse develops.

I'm an artist and I want in. How do I create an NFT or my own collection of them?

Creators and individuals with technical expertise (such as coding in Solidity for Ethereum-based NFTs) can collaborate to make NFTs from scratch. Generated art provides a way for NFT creators to replicate one element into many, displaying varied traits with the help of AI.

But if an artist has digital work already created and can convert these pieces to .jpgs, they can go right to a marketplace like OpenSea to mint this artwork and make it a reality.

Launching or purchasing an NFT of your own could be an entry point into a new community — or several. For many artists, this sense of belonging and kinship is a life-changing result of joining these collectives. Converting your .jpg to an NFT may be the first step to fulfilling a creative pursuit or purpose that's delightfully different from corporate-run, government-sanc-tioned or fiat-driven constraints that are becoming a thing of the past. For others, it's simply about trading, ownership or selling a product. The point is: NFTs serve many different purposes for creators and collectors alike.

The Wrap-Up

From Bored Apes to Weird Whales, we're seeing creativity reinvent itself in a new medium right before our eyes...and it's a beautiful thing.

NFTs redefine ownership in the context of our rapidly changing tech landscape. They can also bring people into the world of crypto, introducing individuals to the possibilities of the blockchain and decentralized currency — much in the way that art and music have traditionally served as a gateway for communities and cultures to define themselves and adopt new practices, beliefs or attitudes.

So the next time someone asks what a .jpg with a price tag can really do, who knows — you might feel compelled to wax poetic about your favorite artist or collectible and the unique value it holds *just* for you.

NOTE: *The comment about the Sotheby's Bored Ape in this intro is an example of recent pricing around a set of NFTs from* *this collection, and is not meant to represent actual up-to-date price info.*

* * *

NON-FUNGIBLE SENTIMENT: EXPLORING THE VALUE OF NFTS

At the beginning of 2021, I had a very narrow vision of what NFTs could be. Their look, their feel, their essence…it all fell under a singular cartoonish umbrella. I had seen pictures of Bored Apes and Crypto Punks floating around, and a couple of my friends had profile pictures that seemed a little less human (and a bit more intriguing) than I was used to.

Don't get me wrong, I thought these images and ideas were cool. But I guess that was just it… they didn't make my blood run hot.

That was until I fell in love. Ensnared by the icy gaze of a blue-tendrilled Crypto Chick, destined to be mine.

Her S-shaped curls? Mine. That layered black top? Wear it all the time. Even the exasperated scowl looked like it could play the part.

So I looked into this particular community on Twitter and tumbled down threads praising women-operated art collectives around the globe minting their artwork or purchasing NFTs on OpenSea. With more digging, I connected with creatives over mutual objectives in fintech — then discovered other marketplaces and creators I'd never heard of before.

This one piece was entirely new to me, except for the fact that it slightly mirrored some of the Build-a-Bear-esque traits of the NFTs my boyfriend had been collecting: mischievous expression, funky outfit, sunset martini backdrop. Layered attributes were laddered together like a colorful recipe. But this chick stood out for a particular reason. She was undeniably femme.

In daylight, she could represent my business — my brand and personality's web presence. I could parade the image as an emblem of my involvement in the crypto community as a writer, strategist and content creator. Under the right conditions, she could represent *me*.

I've been learning a lot about crypto this year — as much as I can wrap my cranium around. But I'll be honest in that NFTs were sort of the final frontier for me of this entire ecosystem. First, there was too much to understand beneath the surface: smart contracts, minting, gas fees. It seemed a fascinating concept, but I questioned if this was all an overhyped fad. If you can "NFT" just about anything, where does it end? The regulation was already way too difficult for me to intuit when it came to Bitcoin, altcoins and exchanges...never mind an ocean of pricey .jpgs. NFTs interested me, but they didn't *captivate* me.

Until I saw this: a minted coincidence met with charming aesthetics and a devilish attitude that held unique value within the communities I cared about.

Once I started using my digital art as a PFP (profile pic), it became intertwined with my online identity. It acted as a stamp of approval for many innovators in the tech space along with Discord groups and YouTube followers. Some of the NFTs I started seeing were not only in line with my aesthetic but helped me push beyond it, growing a taste for the eclectic and a penchant for the digital macabre. At times, I felt like I was part of an underground metalcore club by association and I couldn't help but laugh. This was my first foray into non-fungible artwork, but it certainly wouldn't be my last.

NFTs can be a little unruly. Whether you're creating them or buying and selling, there's an element of intrigue where art collection, hobby, and gambling collide. Being a collector becomes suddenly accessible, personal, and sometimes a little unpredictable.

A non-fungible token is exactly as its name describes: a unique item (documented on the blockchain) that can't be replicated or substituted, and therefore elevates and destabilizes the traditional meaning of ownership. It's also a bit of a rabbit hole where everyone's allowed to be a little mad and encouraged to express themselves in artful ways.

And that's kind of the whole point. The value of creation isn't set by strict rules dictated by some central art authority; it evolves organically within an entire global community. Which is exactly why this cerulean lady meant something to me — whereas to another person, it might just be another one out of 10,000. Or maybe they would find it just as incredible or aesthetically impressive for a different reason.

From certificates of educational achievement (insert cheesy

proof-of-work pun) to video games and real estate, NFTs can cover a lot of ground. There's something for just about everyone to bond over. And if you don't see anything out there that speaks to you? Create something that does.

Of course, there's the argument that trend monetization makes some NFTs shopworn — like a passing fad, it could be a flash in the pan...and more likely someone's short-term gain than "real" art. But can you ever put a price tag on a network?

Cryptocurrency, and blockchain at its heart, was founded on the idea of connecting people without a central authority to do it. Ledgers are a framework for individuals to come together authentically and exchange value through trust *and* proof of the work that established it. Each transaction is recorded in good faith — it can't be changed or moved.

This is blockchain technology's greatest strength, and it's no small wonder we see extensions of it from Bitcoin all the way to its use in the arts. Collection, ownership and purchasing power are the fabric of the social bonds that surround NFTs. Uniqueness, humor and beauty are what give memes, .jpgs and innovative music tech their promise — extending beyond currency and into the community. There are all different ways humans determine worth, but there is something most unique about the connections between people that ultimately decide what is valuable.

Reminder: *At the time of writing and posting this piece online and publishing this collection of newsletters, I own the Crypto Chick NFT mentioned. Nothing within this piece or throughout this publication is financial or investing advice. If you are interested in doing any investing of your own, you should conduct your own research and not rely on anything mentioned here.*

* * *

SHOPPING FOR NFTS: MAKING A LIST

Whenever I'm shopping, especially for anything that's super in-demand, I find it helps to have a strategy in place before setting foot in the store or adding things to my cart. It's the same with NFTs. Everything about that purchase is up to the individual — and I won't lie, I like to do a little soul searching before plunking down any currency on an item...no matter how digital, fungible, niche, trending or tempting.

This is the first of two chapters that will detail the purchasing process when it comes to non-fungible tokens. While I won't be giving you any financial or investing advice, I can share some insights on the where, when and how of the NFT marketplace. But first, I'd like to start with some topics that I consider when I'm thinking about placing a bid or adding to my collection.

Checking It Twice

Here are a handful of questions I explored before buying my first NFT. I revisit this list often.

1. What is the scarcity / rarity of this item and is that important to me?

In the NFT world, scarcity is a key concept. It refers to the availability or exclusivity of the item. It typically goes hand-in-hand with uniqueness, which is often based on the composition of traits displayed in a given token. When someone is looking to trade their NFT, scarcity may play a role in how much a buyer bids — helping to determine its value. Of course, this ranking may not matter to you, and you may base your decision on several other factors.

Rarity guides can serve as a great tool for determining the rarity of an item in a particular collection before you buy, as they help to give information on the traits of various pieces.

2. What's my relationship to the hype? Is this my hype or someone else's?

Sometimes there's an overlap between hype and

rarity. But what I really mean is the promotion and the build-up, or the cultural zeitgeist — how do *you* feel about it? What might your relationship to this item, trend, aesthetic, celebrity or collection as a whole look like weeks, months or more after purchasing? This can have implications based on how long you're aiming to keep the item *and* why you're buying it in the first place. Which leads me to the next point...

3. Why do I want it?

Is this an investment piece? Art that you can't wait to stare at every morning? Is it an avatar for your profile picture or your ticket to an event? You may not have the answer to this question before buying, and there's nothing wrong with that. What's more, you might think you have the answer only to later change your mind. But I put it out there because the question is still worth asking. You never know; reflecting on this topic may help make the buying or selling process a little clearer.

4. What digital rights, if any, come with the ownership of this NFT?

Property rights and trademarks are something that vary from NFT to NFT. I always like to be aware of the rights surrounding the item I'm purchasing; it can help to shape context around what I'm allowed to do with it and how I may want to use it in the future.

There are no right or wrong answers here. Zero judgment — it just comes down to where you derive value and how a bit of thought may help shape your process. Now that we've laid the groundwork for the motivations and goals behind NFT ownership and trading, let's get into how and where to buy them.

* * *

SHOPPING FOR NFTS: SECURING THE .JPG

Let's continue the conversation around NFT drops, bidding and buying. You may have come here for the basics: the where and the how. Got you covered there. While I can't give you investment or buying advice, I can tell you a bit about what's out there and give you (let's go with an obvious analogy) directions to the mall.

You Know the Saying: Phone. Keys. Wallet.

Much like heading out to a physical store, you probably

wouldn't want to forget your wallet. With digital art and collectibles, setting up a wallet is the first step to ensure you're ready for a purchase.

There are different wallet options, and you might choose one based on which blockchain supports a specific NFT and which marketplace is selling it. Some examples are MetaMask, Trust Wallet, and Coinbase Wallet (of the crypto exchange). Wallets can serve to not only hold the funds used to buy the NFT, but as storage for the NFT itself.

Note: sometimes it can take a little while to see the money transferred to a wallet show up in it.

In the interest of time, I am not going to detail how to set up your wallet...but, as always, do your best homework through resources that you trust. Whether it's an educational podcast or a trusted friend or mentor, a little guidance or practice can help with the learning curve that comes with moving crypto around.

Once you have a wallet (or several) set up, there are different methods of copping your digital art. Some NFTs are sold at a fixed price, while others are auctioned off for a set duration of time.

A smaller number of marketplaces take a debit or credit card transaction, if you want to pay in USD/fiat instead of purchasing or moving your cryptocurrency. These tend to be smaller, more niche drops. An example would be Nifty's (different from Nifty Gateway).

Which Way to the Mall?

A few popular marketplaces include:

OpenSea

Supports many payment types and allows artists to create and mint their own NFTs on site, sold in ETH. One of the most popular options with a wide assortment of collections.

Rarible

Multi-chain (ETH, Flow and Tezos) platform for creating, selling and buying NFTs. Features coveted art from creators like Pak, LIRONA and eBoy. Rarible has a $274M trading volume with 1.6M users at the time of writing this newsletter. It connects to MetaMask wallet and has its own RARI token. They've also opened the platform to metaverse land, wearables, domains and more.

Axie Marketplace

Online shop for Axie Infinity gaming. If you've been hearing buzz about the metaverse, know that these two go hand-in-hand. In this colorful world, users can explore in-game NFT experiences and play-to-earn opportunities.

Nifty Gateway

Digital art platform and auction "house" for some of the most popular NFT artists like Beeple and LOGIK. Acquired by the Winklevoss Twins (owned by Gemini). This is a custodial marketplace, so your NFTs are stored on the platform itself.

SuperRare

White-labeled marketplace for NFT creators and buyers, accepting Ethereum. They have their own ETH-based token ($RARE).

NBA TopShot

Associated with the NBA and WNBA. Uses the Flow blockchain to bring elite names and sport fandom to the world of crypto collectibles.

Atomic Hub

Allows users to explore, create, trade or sell NFTs on the WAX (Worldwide Asset eXchange) blockchain. Some of these items can be used in play-to-earn games on WAX as well.

There are others, so please note this is not a complete list... but hopefully enough to get you familiar with the environment.

Waiting in the Queue vs. Window Shopping

Last note: some NFT drops require a lot of patience. This may mean waiting in line with thousands of other people to purchase the NFT you want, often receiving a number assigned to you (like at a deli counter, but let's hope it's more exciting) and having an NFT picked for you at random once your number comes up.

If you're purchasing from a marketplace like OpenSea, the shopping experience is a bit more traditional and you can curate it to your liking — browsing various collections and exploring what makes them unique or appealing. There are typically dropdowns to help narrow NFTs by certain traits.

How Do I Know if the Bag's Designer?

On OpenSea, a blue check mark indicates that the seller has been validated (like an accreditation, ensuring their identity is valid and the collection is authentic). This is a helpful tool if you're unsure of who is behind a specific account or set of NFTs.

I like to double-check by locating the account on Twitter (X) or surfing the web for info about the artist or creator, too, since I enjoy learning the story behind the artwork I purchase. For a lot of people, the clout is what clinches the sale — but for others it's more about the come up.

I hope this was helpful, and that you may come back to visit this chapter on occasion when you're exploring the vast world of NFTs or explaining them to a friend who's interested in doing the same.

* * *

DÉJÀ VU WITH GOGGLES ON

Recently, people have been using the term metaverse to imply a new idea. Truthfully, we've been spending time in virtual places for years.

Gaming, especially, has gifted us these types of escapes — from the Sims to Grand Theft Auto. Since the late 1990s, we've sort of been participating in mini, beta metaverse realities...

...leading me to the question: *is this déjà vu with goggles on?*

I'm reaching, but the point stands: there are threads of today's VR experiences that date back decades. The main difference is that we're seeing extensions of these same general themes blossom as creators take advantage of all the technology we have today. So yes, the goggles and fancy hardware do have something to do with it.

Companies and developers are making gaming realities *interoperable* to create an immersive, digital shared space that exists on and across blockchains.

I'd like to highlight some blockchains and tokens related to the metaverse: a network of virtual spaces where users, builders and AI can interact. This can include virtual and augmented reality — and can serve as a venue for gaming, socializing and earning.

It might help to set some guardrails with a quick vocab lesson.

The Meta-Glossary: What Is Extended Reality?

Virtual reality, as we're used to hearing a lot about, is commonly defined as a first-person, simulated experience or environment. It may bring to mind classroom learning tools or several brands of headsets that provide realistic depth of vision. There are a couple other terms that you may find helpful in describing the metaverse experience.

Augmented reality is described by VentureBeat, somewhat similarly, as an experience in which "virtual content is spatially registered to the real world...providing a strong sense of presence in a combined real/virtual space".[31] Some examples of AR include Snapchat, Pokémon Go, Google Glass and interior design apps.

Another term that's often thrown around here is basically a combination of both VR and AR spaces: *extended reality*

(XR). XR is a term used to encompass all of what VR and AR can do. XR exists at the intersection of real world and simulation. *Mixed reality* is yet another term commonly used to express the synthesis of real and virtual or computer-based environments.

Now that we've covered what the metaverse is all about, let's build a bridge between crypto "altcoins" and the worlds of gaming and XR.

It is with excitement, and a little trepidation, that I write about these next few innovations. On one hand, it's crucial to give a bit of exposure to the metaverse through avenues as decentralized as possible — to shed light on what's happening in this space outside of huge monopolies bringing the buzz to headlines.

With that said, it's important to understand that the scope of metaverse technology is so vast that it extends far beyond cryptocurrency. There's exciting overlap, but what's presented here is only a slice of what's being built in this space. By the time you're reading this, some of the following projects may look quite different.

By writing about some of these metaverse-related cryptos, I'm not singling out coins based on their perceived merit, trading potential or dominance in the market; rather, I'm sampling what innovations in the space represent utility in connection to virtual or extended reality. Just keep in mind: these are only examples, and there's lots more to learn.

Avalanche (AVAX)

AVAX is the native token of the Avalanche platform, an open-source blockchain that operates with smart contracts as a DeFi haven of sorts. They are on the more decentralized end of the spectrum and brand themselves as "blazingly fast, low cost, and eco-friendly". They offer staking opportunities (proof-of-stake) to process transfers and help keep the platform secure.

Polygon ($MATIC)

Polygon is a proof-of-stake multi-chain scaling platform built by and for Web 3.0 developers. This means it provides infrastructure solutions compatible with Ethereum and other blockchains — benefitting from the network effects of Ethereum and helping dApps improve performance. Their primary aim is to solve for issues like high gas fees and slow speeds. $MATIC is the name of their token.

Decentral Games (ICE)

Games, games, baby! Decentral Games has been seen as a

sizzling hot spot for play-to-earn (P2E) gaming — and the opportunities afforded by its in-game currency (ICE) spark global excitement.

ICE poker is one of the games that allows individuals to earn liquidity and participate in their NFT wearable collection. An NFT wearable is required to play — which can be purchased, traded or borrowed. Individuals can vote in the Decentral Games DAO with its governance token (DG) to make changes and issue fees, and guilds allow built-in delegation to structure the community.

Decentraland (MANA)

Decentraland is a virtual reality platform that runs on the Ethereum blockchain. The platform uses tokens called MANA, which individuals can use to purchase plots of land to sell/ trade or build upon (or to buy other virtual goods).

Corporate and Celebrity Adoption

It's nearly impossible to avoid the topic of the metaverse in crypto spheres lately. Institutional adoption has seen a rise within the gaming economy and virtual reality — as in the case of Microsoft buying Activision for $69 billion or the Facebook/Meta rebrand and their Oculus products.

With celebrities like Eva Longoria and Paris Hilton unfolding their own NFT-driven metaverse projects, we're also seeing tech spaces for women blossom with a little nudge from celeb endorsers. Women-run NFT collections like World of Women and Crypto Chicks are forging connections as they prime their communities for XR launches and concepts. In some cases, this even includes 3D avatars.

The Wrap-Up

Metaverse projects, although unique, typically share one thing in common: they aim to provide tools to gamers and creators to earn and advocate for their own skills...and innovation in this space has allowed communities around the world to earn in new, surprising ways.

Of course, with any crypto investment or venture, some of these projects have proved or will prove more successful than others. As with the spirit of gaming, there are wins and there are losses.

Though some of these games and metaverse spaces offer potential liquidity and staking opportunities, all investments come with risk.

It's also worth reminding ourselves that while these altcoins hint at exciting innovation, they occupy very different markets and use cases than Bitcoin. The capabilities of this first, successful digital currency are unparalleled in many ways; Bitcoin is unlike any other crypto in how it may open the world to financial sovereignty. Bitcoin holds potential as a global store of value, with traits that allow it to solve some of the greatest financial challenges of our time. The altcoins here serve different purposes, and although they're fun to learn about, they shouldn't be mistaken as equal.

To sum it up: these creative tech projects serve a different (but complementary) purpose.

We'll have to wait and see whether metaverse enthusiasm increases in years to come, or if all the buzz taking place in 2021—2022 was mostly a result of corporate buy-in. There's lots of speculation in the crypto space that things like NFTs and metaverse products could be a flash in the pan; alternatively, some individuals are banking the moon.

I am hopeful that the more decentralized projects in this space — ones that hold respect for the ethos of gaming and our spirited crypto community — will lead to building safe, fun and accessible experiences for all players (where we can meet again and again).

* * *

DECENTRALIZED METAVERSE
AND THE VR SPECTRUM

[Note: This article was written in April 2022 after the unveiling of Meta's product Horizon Worlds became newsworthy and started to drive headlines.]

Define "Metaverse" for Me One More Time

The metaverse is a virtual network of spaces where creators, gamers and AI can collab. Think of it as one big playground for users to build, socialize and even earn crypto.

Extended reality (XR) is a term coined to encompass all elements of virtual reality (first person, simulated environment) and augmented reality (first-person experience in combined real/virtual space) that we have witnessed over the years in gaming tech. You may become more familiar with this term as it's used to describe several types of metaverse-related play.

Who's Playing?

We've seen some interesting things play out in the first few years where crypto and the metaverse began to collide. And I'm not talking about 3D nude NFT avatars. Although that has been an interesting part of the optics.

More broadly, the notion of decentralization has been

questioned and redefined within this sphere. While the altcoins that fuel the crypto gaming space exist on a spectrum of decentralization, so does the tech and the infrastructure.

We have, on one level, fairly decentralized tech offering opportunities for games and building land within this space. Yet we also see large, centralized companies making moves by offering play experiences, tools and accessories. In theory, this coexistence sounds pretty peaceful. Collaborations have always been part of tech development and adoption; in fact, sometimes they're necessary.

In 2021, something interesting happened. Social empire Meta (formerly Facebook) announced they would focus efforts toward development in the metaverse arena. This was different, because it wasn't a collaboration or small endeavor: Facebook gave their entire brand a new name and roadmap around the potential growth of this space. Here we have a very structured company, with lots of monetary success, incentives and centralized control, creating an identity around a community that defines itself as roughly the opposite. Some immediately dismissed this as a bad idea, while others were excited about the potential innovation. For a lot of people, the ideas just didn't match up, which left them feeling confused.

Hopefully Nothing?

In 2022, Facebook/Meta began headlining crypto news after unveiling plans for their VR game Horizon Worlds. But with a cut of up to 47.5% on in-app sales, critics were questioning whether this is an oversized "byte" of the pie.

It was more than Apple charges devs for in-app purchases, which Zuckerberg had gone on record saying he'd like to compete against.[32] For more decentralized metaverses like The Sandbox and Decentraland, in-app fees can look more like 5% or less, plus any gas fees.[33]

Of course, each platform has unique attributes, and elements like in-game NFTs and play-to-earn experiences all play a role in what a developer or user might potentially make or spend while building and playing...in any metaverse reality. So keep in mind that these estimates vary depending on the platform and each unique experience.

Both Sides of the Coin

After Meta's big rebrand in October 2022, we're seeing them make very real moves into the metaverse with their new avatar-based app. Plenty of gamers are thrilled by the VR experiences Facebook's Meta Quest (Oculus) products have to offer, looking forward to any novelty that accompanies the brand shift along with refreshing ways to use or expand on this tech. For many, it's exciting to see what Meta is planning and to confirm that they are, in fact, working on something very tangible.

However, all this buzz doesn't come without speculation and concern. The initial rebrand last October had the crypto community in feisty debate over Meta's anticipated role in the space, and now conversations surrounding this rollout have turned up the volume.

The move has sparked discussion on what the future of VR tech will look like with big names — especially centralized ones with lots of money and, in this case, a controversial history around data and privacy — joining the space.

Etiquette and Expectations in Virtual Reality

Here's a key line of questioning that I believe this all boils down to:

In a world where innovation is so expansive that almost anything can be dreamt up and implemented — a new landscape so vast that the goal is to create, inspire and expand without traditional limitations — how are companies and users both expected to behave?

What are the rules...are there any? And is there a shared moral compass when navigating this creative expanse?

While the topic at hand hinges on money morals, the implications are more far-reaching. When we think about tech companies acting ethically, it brings to mind issues of privacy, inclusivity and safety. This is what worries me most when conceiving of a gamescape where reality is intentionally blurred and identity can be as shadowy as a cipher.

At the same time, if we are to innovate and allow for unbridled imagination, there's less of a rulebook for crypto than exists in traditional tech and e-commerce spheres. With greater decentralization comes increased opportunity for budding ingenuity, even though the space is a bit uncharted.

Maybe we're betting on a new way of doing things, one that's naturally incentivized to act on behalf of consumers' best interests when the nucleus of authority is removed from the equation.

I would hope, of course, that most innovators and players in the space aim to incorporate both: a radical approach to creativity along with deep ethical decision-making, driven by the collective intellect. Time will tell, and we'll be keeping an eye on this space.

PART V

Resources for Your Career in Crypto

CHOOSE YOUR ADVENTURE:
BITCOIN AND CRYPTO CLASSES FOR EVERYONE

We often get asked about crypto classes — how we first learned about Bitcoin, what courses are the best, how to connect with other learners — so we would love to spend this segment outlining our thoughts on education in this space.

After all, this is not only how we both started the newsletter, but also how we launched our careers: getting to know the ins and outs of Bitcoin with Anthony Pompliano and his team.

Whether it's a formal online program with a built-in social component or a YouTube series to watch in your downtime, there's so much quality content out there to explore.

Even the bear market blues provide the perfect setting to get back to your roots in crypto or start learning about Bitcoin for the first time.

The Crypto Academy

We are biased because this is where we got our start in crypto, along with the opportunity to meet people all over the world.

Led by Anthony Pompliano and a handful of coaches, the course spans several topics. Students come from all different backgrounds and levels of Bitcoin knowledge. Breakout groups and assignments provide room to explore. There's even a career day for students to meet crypto company reps, and a job board that's open to everyone.

He dips into topics like Web 3.0, smart contracts and Ethereum, but Pomp's curriculum mainly focuses on Bitcoin's technical foundations and purpose — so don't expect to be learning about Dogecoin and chatting about alts all day...

MIT Open Courseware

Gary Gensler leads this series (which you can find on YouTube). It's a sweeping overview of cryptocurrency and the role digital cash plays in our world.

Learning from the chairman of the SEC certainly adds to the prestige, but these classes aren't stuffy in their approach. From ledgers to smart contracts to regulation and some topics beyond Bitcoin — there's a bit of everything, and the classes dive deep into each.

It feels like you're sitting in on an MIT lecture, and as of writing this, it's free. This open course model is one that I'd love to see more of when it comes to higher education in America, in fintech or any other field. This course was recorded in 2018, so keep that in mind if you want up-to-date topics only.

Of course, you can also enroll in MIT's short course (or Wharton's), but it's got a price tag and, to be honest, I don't know how much more it offers compared to the free YouTube classes. I would guess the benefit to enrolling might be the community aspect and learning from live lectures or practical exercises.

Fintech Bootcamps

Berkeley

Intensive courses at top-tier universities like Berkeley offer technical and hands-on learning opportunities — where students can earn a certificate without enrolling in a standard, full-time Masters program. This environment could serve as a great option for learners looking to shift lanes in their career without taking significant time off.

For folks who don't mind the financial commitment, I've heard bootcamps can be super instrumental in allowing learners to absorb a ton of information in a timeframe that suits their lifestyles and career goals. Plus, this one's a mix of crypto, coding and general fintech...the perfect combo for anyone looking to make a switch within their industry or change tracks altogether.

UMiami

This 24-week option seems similar to the Berkeley bootcamp mentioned above. The creators of the course aim to prep participants for a career in fintech through practical projects and technical training. The curriculum allows learners to experiment with analysis tools, machine learning algorithms and programming languages like Python. It sounds to me like these bootcamps all have some similarities: sprawling workshop topics, impressive accolades, necessary self-direction and a dash of "you get out what you put in".

Coursera

Copenhagen Business School: Digitization of Finance

I thought this series would highlight crypto, but there wasn't a ton of that; it was mostly rooted in business. I'm so glad I took the course, though, because I didn't have that background beforehand. If you've been through business school or launched your own start-up in the digital age, then I would honestly say you can skip this.

But for individuals looking to learn about the basics of digital transformation in banks and business, perhaps in order to lay out a case for Bitcoin or understand the integration of digital payments across industries, this could be the class for you.

One note: Coursera is very much a DIY experience. You will need to pace your work and more or less instruct yourself at times. You likely won't get the same sense of community as you would with breakout groups or conference calls, and the quizzes and projects are a unique blend of open-book and just-roll-with-it faux homework.

Since the instructors are multilingual, there are times when American learners may feel a divide in terms of the content and instruction — but that step outside my traditional comfort zone was a good thing. Any course that offers exposure to nonnative learning or collaboration with classmates from overseas provides value, especially when it comes to a globally significant topic like fintech.

Stanford University: Cryptography

This computer science class brings it all back to the deepest roots of cryptocurrency: privacy and encryption. It starts off fast. I got through a chunk of lectures and then had to take a break when my freelance work picked up...but it was so cool, and it got me hooked on solving ciphers.

For the first couple of lessons, it seemed like the instructor was writing in hieroglyphs all over a whiteboard. Like Good Will Hunting, only pre-recorded from the West Coast.

Do you love 90s grunge and dark academia? Are you a number theory nerd or algorithm curious? This could be a good choice for you.

The work was worth it, ultimately, because I was able to tap into an old, cherished love of secret codes and real-life puzzle sleuthing — something I've romanticized since I was a kid. If that sounds like you...do your thing, Dan Brown.

Whether you're just getting started or still learning, there's a crypto track for everyone, and so much content online aside from this list. If nothing here sounded quite right, DYOR and I'm sure you'll land on options suited to you.

* * *

THE NEW BITCOINER'S BOOK LIST

Plenty of fantastic crypto education exists in bite-sized blocks online, but books will always be one of the best learning resources (we're biased).

Before making any recommendations, though, I wanted to get several 5-star reads under my belt. I spent a good chunk of 2021 and 2022 with some of the most interesting texts on cryptocurrency and our global economy, and I'm thrilled to share my top seven with you.

What I Recommend for Newbs — and In What Order

These are all books that I've read (or listened to on audio) and discussed at length with peers. While there are many other titles out there that could surely compete for these spots, we'll stick to the stuff I know so I can give an informed synopsis.

These will appear in a particular order: readings I see as the most necessary for newbies *as well as* what might complement each other best when read one after the other. That way, these titles can help build your Bitcoin knowledge (in an order that makes sense) as you work your way down the list. We will start with my strongest reco in the #1 spot, and I'll indicate whenever an alternate reading track may be preferred for a given audience.

Happy reading!

1. *The Sovereign Individual*, James Dale Davidson and William Rees-Mogg

A crucial read for any member of the Bitcoin community, *The Sovereign Individual* outlines our shift as a society into the information age. It was published in the 90s, and so much of what the authors imagined the world to look like now has already become a reality.

The last chapters detail what it takes to embrace a sovereign lifestyle and become more independently prepared for social and political changes. After all, if institutional power starts to destabilize, those entities would likely try to reclaim their strength. Old customs might perish, while familiar historical patterns reappear. Confronting the new paradigm, strong individuals must adopt new practices.

Spoiler: the ability to travel and explore outside one's comfort zone — plus owning a unique set of skills suited to flex with global needs — might allow a person more comfort and stability within our shifting digital landscape.

Self-reliance and freedom of thought are paramount, as tech advancement prompts us to rethink constructs of *value*.

The Sovereign Individual takes this #1 slot because it sets the stage for why Bitcoin is so important. However, it may also be effective to read this book *after* gaining a solid introduction to how Bitcoin operates (see #2). Either way, it's necessary reading.

It helped me envision how digital currency ties into society, culture and our future as a species. Thanks to this foundation, I am more confidently constructing and refining my role as an individual within the context of the macro picture.

2. *The Bitcoin Standard*, Saifedean Ammous

The Bitcoin Standard is often considered the holy grail of Bitcoin texts due to its organized chapters, comprehensive guides, and insights from an author who deeply understands the nuances of digital currency. Saifedean covers topics like the history and origins of money, its role within our global economy, plus how Bitcoin solves for specific financial and sociopolitical obstacles.

Note: It does get into some technical nitty-gritty, so if you're looking for a lighter read with a more sweeping financial lens, my #5 pick (*The Price of Tomorrow*) might be a more suitable intro for you. That said, if you're looking for a Bitcoin focus... stick with Saifedean!

EXTRA CREDIT: *The Fiat Standard* — Read Saifedean's most recent work to learn more about our banking system and how debt and inflation are tied to the minting of fiat currency. In my opinion, Saifedean's reads are best treated as a package deal; these two pair perfectly in succession, or even at the same time.

3. *The Book of Satoshi*, Phil Champagne

This book is so cool; it felt like I was reading the diary of Satoshi himself. There's correspondence between the anonymous creator of Bitcoin and his colleague Hal Finney, and each chapter begins with a summary of the conversations that follow.

You'll feel like you're eavesdropping on chat forums with Satoshi and the earliest players in the crypto space, just as Bitcoin was first emerging. It's one of the best ways to learn about the ins and outs of Bitcoin and the exciting journey these brilliant minds experienced when creating and implementing the first successful digital currency.

Phil Champagne is a fantastic writer and curator with keen insights into this thrilling block of time when Bitcoin was in its nascent stages. You could take on this collection at any point in your Bitcoin journey, but having some basic knowledge beforehand doesn't hurt...so I'm placing it after *The Bitcoin Standard* because of its level of nuance. I encourage readers to brush up on the basics of Bitcoin (as a store of value and medium of exchange) first, and then dig into its historical significance with *The Book of Satoshi*.

To put it simply, this one ranks high because I had the most fun reading it...and I think you will, too! In terms of pure enjoyment, it'd be an easy number one — and I have yet to come across another resource that reaches this level of detail about the creation of Bitcoin. Perfect buddy read or book club pick.

Tip: It's laid out nicely for e-readers, if that format appeals to you. It's easy to skip around if needed, plus lots of quotes serve as ideal pivots for highlights and annotations. I would also highly recommend this author's second book, *Bitcoin vs. Altcoins*, for a deeper dive into what sets certain cryptocurrencies apart from the rest.

4. *21 Lessons*, Gigi

Gigi is a well-respected engineer and writer in the Bitcoin space. He worked on the Bitcoin code and has penned *21 Lessons: What I've Learned from Falling Down the Bitcoin Rabbit Hole*, a series of essays on cryptocurrency and the history, function and potential of Bitcoin. I would highly recommend this for anyone first starting out in the space because the topics are far-reaching — but crucial to understanding this tech — and he breaks them down eloquently.

If you're considering an alternative reading track, this may be a good first step for those who want the lowdown on crypto but feel they're not yet ready for Saifedean.

You can check out the Der Gigi website (dergigi. com) to read or listen to some of these thoughts in any order you please. When I first began taking Bitcoin courses, I listened to a few of these audio recordings on-site, and they helped me grasp so many key concepts that I was struggling to understand in class or conversations with my peers. His "Bitcoin is Time" may be my favorite of the bunch (prepare to have your mind blown).

The next three recommendations will cover titles with more of a macroeconomic focus, emphasizing financial crises and solutions on an individual or collective scale. They'll touch on the factors that could lead to monetary, political and social infrastructure collapse or devaluation. Sounds scary; Bitcoin can make it better. As we say in the crypto community, "WAGMI" — we're all gonna make it!

5. *The Price of Tomorrow: Why Deflation is the Key to an Abundant Future*, Jeff Booth

I'd think of this book as the optimistic-yet-practical, entrepreneurial sibling of *The Sovereign*

Individual. It, too, delves into our changing technological world — while placing a stronger emphasis on collective industry and innovation as a hedge against debt and inflation. Booth writes about efficiency in tech and how deflation can change the course of a system relying far too much on antiquated measures.

This book would also pair wonderfully with The Fiat Standard for anyone looking to dig deeper into the trappings of our current monetary infrastructure and the future of our global economy. Both together would make a lovely gift for someone who is passionate about these topics.

The takeaway: If you're looking for a book that zooms out on the practical application of Bitcoin — start here! Perhaps you are a student of economics or starting your own business and interested in the impact of crypto? This is your #1.

6. *Do Bitcoin: The future of money. And what you need to know*, Angelo Morgan-Somers

This book takes a breezy, uncomplicated approach to explaining Bitcoin's purpose and value. Angelo is great at distilling complex topics into lessons that are enjoyable to learn, and clear sections make it easy to share your new knowledge with others. This work covers necessary Bitcoin and blockchain topics in a relatable voice without overwhelming the reader. *Do Bitcoin* is an ideal companion for beginners (and for more advanced readers seeking a refresher) — and it makes a great gift, too!

7. *Principles for Dealing with the Changing World Order*, Ray Dalio

This read will likely split a lot of crypto enthusiasts down the line in terms of whether they'd praise or pass. Even though it wasn't as scintillating for me as, say, *The Sovereign Individual* or *The Book of Satoshi*, I can see why it's a key talking point for so many Bitcoiners.

Dalio touches on open-mindedness, weighing risk and reward in entrepreneurship or investing, and the concept of an idea meritocracy. This book focuses on issues cropping up in the Western world and how he believes we can best position ourselves in light of these changes.

I could do without the self-help vibes, but there's a

good chunk of his argument that hits right where crypto delivers. So while it's not the type of book I usually lean towards, it *is* one you might find yourself discussing with friends or colleagues in Bitcoin and finance circles. And because those encounters could be the most significant ones you have in your career...it's on our list.

Dalio emphasizes what brought him success throughout his career and why some of these traits are crucial to thriving in the information age. One of these truth nuggets is the importance of listening to and learning from people who challenge your preconceptions — something most truly innovative leaders would likely agree on.

The Wrap-Up

Taking a look at the full list, a few of these works are more foundational to the understanding of Bitcoin and what it solves for, while others paint a more general picture of our financial and social landscape. All of them, in my opinion, can shape a more comprehensive understanding of digital currency and how we might prepare ourselves for global financial challenges.

* * *

FIVE WAYS TO STAY
CREATIVELY INSPIRED IN CRYPTO

Every industry requires innovators. Within all of us, there's a capacity to flourish and fill these roles with hope and excitement. Yet during this unique Great Resignation, workers are demanding more — and the reasons for this shift aren't all that rosy. From pay gaps and glass ceilings to childcare challenges and inflation, the global demand for change is not about to trade its long-term agenda for a quick buck.

We've seen crypto and fintech ecosystems play a huge role in redirecting time and talent towards a digital focus. Whether they're Bitcoin enthusiasts starting new roles outside of legacy corporations, day traders taking their side gig to the main stage or working moms balancing online classes on their already full plate...crypto has carved out a niche for those looking to blend old cash rules with a little novelty.

I'd love to tell you five different ways that I've come to terms with my own "great resignation" by integrating Bitcoin and crypto into daily work. While a couple of them actually did revolutionize my career, the rest just made it a little more fun.

1. Find Your Tribe

For me, it was a Bitcoin class, Discord channels, Twitter Spaces and a NFT poetry group on Instagram that inspired and held my interest in the blockchain. Those communities made me realize that despite all the time I spend cooped up working on projects at home on the computer, there are so many open-minded individuals and groups out there to network with. The crypto ecosystem is brimming with innovation on every app and each channel — it's all about tuning in, staying safe and connecting authentically. When you find people you trust, they just might show you a world of crypto knowledge and exposure beyond your own.

2. Decentralize Your Work

We've been accustomed to Web 2.0 for some time now. Whether you're a writer, a designer or any other creative type, you probably rely on some form of social media for work and collaboration. The advent of a new World Wide Web brings with it newer, decentralized platforms that can be a lot of fun to explore and learn from no matter your industry or background.

While Medium, Instagram and Facebook are familiar territory to many, more decentralized products and innovative spaces (Mirror or Minds, for example) could give creatives financial incentives and a whole new audience or community. They're often censorship resistant, which can have important privacy and innovation implications for your work. Ultimately, the key is to reflect on what values you're looking for in a platform, be that centralized or less so — and use those findings to guide where and how you flourish as a creator.

3. Make AI Friends

When first introduced to AI solutions for creative work, I hesitated to use them. But the more I explore these tools, the more I see how they can complement my work rather than compete with it. In NFTs, we see direct and fruitful collaboration between designers and generative art tools. Popular platforms for creating AI artwork include Midjourney or Jasper (and there are many others to check out).

In my early 30s, I encountered a major career

crossroads and decided to give the new tech a shot. In the hopes of booking some crypto content creation gigs, I tried out an AI logo generator to help with the launch of my own brand. At first, I was against everything this stood for. I collaborate with designers on lots of projects, and bypassing this talent or skimping out on a programmed alternative felt cheap. What I learned was that AI didn't need to (and could never) replace human design; it could serve as a step in between. I could use it to generate initial ideas or expand the scope of my concepting stage by initiating keywords, drafts and sketches.

AI didn't prohibit me from working with designers; it just added to what we were creating together by giving me a place to start.

Keep in mind, though, that lots of people feel all different ways about incorporating these tools into their crypto workstreams...so if you are collaborating with another person or company and planning to include AI, it's always worth it to mention this ahead of time and confirm their comfort level.

AI writing assistants can help us turn ideas into full-fledged content, aggregate research behind a concept, or generate a prompt or two to get the ball rolling. This can be an efficient way to start and scale a Bitcoin-centered business or editorial project. We can sync up with AI to streamline planning, editing and presenting, too.

With a bit of learning, some stock and crypto traders are even using AI bots to help save time and hone their strategy (after vetting the right platforms and resources first). When used wisely and transparently, AI is an increasingly important tool to spark creativity and meet the demands of our ever-changing technological landscape.

4. Keep Learning

I've been very vocal about what launched my career in crypto currency: remote coursework and videos on YouTube. It doesn't have to cost a lot of money to learn about crypto, fintech and Web 3.0. Even though we are busy, the accessibility of these classes makes them easier to fit into our schedules. No matter what field you work in or what career shift you're planning to make, there is something out there for everyone — and it just might turn your day job into something you love, full time.

5. Earn Passive Income (While Doing More of What You Enjoy)

Sometimes the art or the creation doesn't bring in a ton of money, and that's okay. With day trading tools, NFT collecting and P2E gaming, there are ways to (cautiously!) explore making some side income through crypto and maybe give you more time back in the long run. As with all of DeFi, there are risks and there's a learning curve — the crypto space is notorious for its volatility — but if that's something you're comfortable with, more and more opportunities for trading, staking crypto or earning yield are available. This is *neither* of our specialties, so we're going to move on from it rather quickly... but for people comfortable with a lot of financial risk, crypto can open the door to certain oppor-tunities. As always, do your own research before engaging in any of these ways to earn.

Now onto something we are *super* comfortable with: newsletters, blogs and self-publishing endeavors serve as a fantastic place for writers to start when embarking on a new crypto career. These activities can also extend a person's network as they learn about Bitcoin. We've found that employers love to see initiative in any form, whatever your passion. Maybe that's creating video content, learning curriculum, a book club or a social media platform for likeminded crypto enthusiasts to connect. No matter your interest, there are plenty of spaces for you to explore and plug into your new excitement around digital currency or Web 3.0.

By joining a DAO, earning through play-to-earn games, setting up your own Bitcoin node or connecting your platform to Lightning Network payments and Bitcoin tips, anyone can begin to monetize these passions in a way that can earn them a few sats or airdropped goodies for all that time spent on extracurriculars.

As we incorporate more creative tools and outlets into our routines, there's room to shift our work-life balance to reflect our authentic desires and passions. Even in unbearably bearish markets, there are lots of exciting ways to stay inspired while learning and earning together.

Final Wrap-Up

Crypto casts a wide net, and we're always finding out something new. The world of fintech will evolve beyond the publication of these newsletters, and we are thrilled to see it grow. With that in mind, part of what's relevant now may change as the years go by. We love that this book may serve as a bit of a time capsule for a few brief but exciting moments of Bitcoin's overall evolution. It's been a wild ride so far, and we've loved every second of learning together.

The Bitcoin ethos is all about sovereignty, and a major step in that direction is doing what you feel is right — regardless of what's popular, easy or normative. We believe this extends into career paths and the complex lives of intelligent women all over the globe. Falling down the rabbit hole involves more than getting psyched about a conspiracy or latching onto a trend without continuing the research. It has to do with seeing how a theory or innovation applies to your life and the world at large.

So, follow your calling...and then keep on digging! When you learn something new, share it with others. When you get bored with one topic, find a story or a conversation that leads you to a deeper understanding of another one. Ask yourself *why* it is that crypto intrigues you, *what* decentralization can do for global currency, and *which* topic resonates the most with you from the book you just put down.

We started on this road as a couple of women in crypto, writing and discussing themes that felt relevant to us and our friends. Eventually, this passion seeped into other corners of our lives. We found ways to shape our careers around fintech innovation and pursue our shared enthusiasm for financial freedom — with the hope that women around the world may discover opportunity, within the Bitcoin space and beyond.

All this is to say: when you step outside your comfort zone to chase what intrigues you, the world might open up in ways you never could have anticipated. We're excited for you to discover your calling and visualize the journey ahead. The world of crypto has encountered unexpected moments... some exciting, some trying, some strange. This restlessness, we are most certain, is how all great things begin.

Acknowledgments

Warmest thanks to our families and loved ones for supporting us on this crypto journey and throughout our lives. We would not be where we are today without your love and encouragement.

Our heartfelt appreciation to Rachael Churchill for editing and proofreading this entire collection. This was a crucial step in bringing our newsletters to print, allowing our readership to grow and explore this content more fully. Thank you for your knowledge, patience and professionalism! We're thrilled your expertise saw this project to fruition.

Of course, thank you to all readers — and especially early followers — of our original newsletter. From the bottom of our hearts, we appreciate your continued readership and early support in our experimental writing project. You're the foundation of this community, and a big reason why we took these career risks in the first place. We've got so much gratitude for you!

Thank you to our newsletter guests for sharing your contributions to our original Substack platform. From guest posts to interviews, your career highlights and crypto achievements cast a broad lens on Bitcoin, digital art and crypto wallet storage. (Readers: you can find these guest pieces on womxnincrypto.substack.com).

Chris Mahadeo, thank you for using your social media platform to teach artists about NFTs and AI; it was a proud opportunity to dive deeper into this tech and explore ways to blend poetry with crypto. We admire your ability to balance the two in new, exciting ways while teaching others.

Pomp and The Crypto Academy's incredible coaches, your class changed the way we look at the world of Bitcoin and cryptocurrency, and stirred us to create something valuable with our time. Thank you to everyone in our cohort for making the learning adventure so fun and inspiring our side hustle.

A big thanks to the entire team at Substack. This platform's user-friendly design and flexible experience gave us the tools we needed to build a community.

Next, thank you to the developers and professionals behind HackerNoon, another site where some of these articles were published. In working with your platform, we found the first truly open and deeply innovative online publishing interface for tech topics and beyond.

Also, thanks to Reedsy for providing an entire toolbox of resources, from editing and proofing to design and layout — plus, helpful tips along the way.

And most importantly...you! We're tremendously grateful for your support. If you found this book at all helpful, or learned an interesting new fact, we would love if you could leave us a review anyplace this book is sold. We wish you the best of luck in your career and blockchain journey.

Many, many thanks!

Author's Note

Some pieces were published on outside platform HackerNoon by Stevie Sats and later included within this collection. This selection includes:

"Edgar Allan Poe Was Bullish on Crypto"
"Non-Fungible Sentiment"
"Deja Vu with Goggles On"
"Decentralized Metaverse and the VR Spectrum".

"Deja Vu with Goggles On" and "Decentralized Metaverse and the VR Spectrum" were originally published on HackerNoon. com under the titles "Have We Met(A) Before?" and "Meta: Horizon Worlds to Take 47.5% Cut of In-App Sales from Game Developers." Some changes to content have been made for the purpose of this publication.

Other posts listed above were first published on the Substack. com platform and deployed as newsletters.

Sources

1. Trust, but Verify. (2023, May 24). In Wikipedia. https://en.wikipedia.org/wiki/Trust,_but_verify

2. Christmas Story (1983)—*IMDb*. (n.d.). IMDb. https://www.imdb.com/title/tt0085334

3. Surz, R. (n.d.). Money Printing and Inflation: COVID, Cryptocurrencies and More. *Nasdaq*. https://www.nasdaq.com/articles/money-printing-and-inflation%3A-covid-cryptocurrencies-and-more

4. Alden, L. (2023, January 13). Bitcoin's Energy Usage Isn't a Problem. Here's Why. *Lyn Alden*. https://www.lynalden.com/bitcoin-energy

5. Channel 4 News. (2018, June 15). *Jaron Lanier interview on how social media ruins your life* [Video]. YouTube. https://www.youtube.com/watch?v=kc_Jq42Og7Q

6. Sherman, M. (2021, September 14). Nodes on Bitcoin's Lightning Network Double in 3 Months. *CoinDesk*. https://www.coindesk.com/tech/2021/07/14/nodes-on-bitcoins-lightning-network-double-in-3-months

7. Bitcoin—Open source P2P money. (n.d.). https://bitcoin.org

8. Lightning Network Statistics | 1ML. (n.d.). Lightning Network Search and Analysis Engine—Bitcoin Mainnet. https://1ml.com/statistics

9. ARK Invest. (2021, July 21). *The Ɓ Word | Live with Cathie Wood, Jack Dorsey, & Elon Musk* [Video]. YouTube. https://www.youtube.com/watch?v=Zwx_7XAJ3p0

10. What Bitcoin Did. (2021, September 3). *Bitcoin Risk Assessment with Lyn Alden* [Video]. YouTube. https://www.youtube.com/watch?v=lyI1jTemwIQ

11. Bitcoin Magazine. (2022, April 7). *Bitcoin 2022 Conference—MAIN LIVESTREAM—General Admission Day 1* [Video]. YouTube. https://www.youtube.com/watch?v=l6YZrCyhmlA

12. Leutwyler, K. (2000, November 3). A Cipher from Poe Solved at Last. *Scientific American*. Retrieved November 1, 2021. https://www.scientificamerican.com/article/a-cipher-from-poe-solved.

13. Morelli, R. (2018, May 3). Edgar Allen Poe and Cryptography. *Edgar Allan Poe and Cryptography*. http://www.cs.trincoll.edu//~crypto/historical/poe.html.

14. Pelling, N. (2008, May 13). "The E. A. Poe Cryptographic Challenge"... *Cipher Mysteries*. Retrieved November 1, 2021. https://ciphermysteries.com/2008/05/13/the-e-a-poe-cryptographic-challenge

15. Seth. (2011, November 1). Edgar Allan Poe Cypher. *Crypto Crap*. https://cryptocrap.blogspot.com/2011/11/edgar-allen-poe-cipher.html#comment-form

16. Ethereum. (2014, July 29). *Vitalik Buterin explains Ethereum* [Video]. YouTube. https://www.youtube.com/watch?v=TDGq4aeevgY

17. Phillips, D. (2021, April 16). What is Near Protocol? *Decrypt*. https://decrypt.co/resources/what-is-near-protocol

18. Ramstein, W. (2022, February 14). Solana too ambitious for Ethereum, targets Nasdaq + Visa. *FlowBank*. https://www.flowbank.com/en/learning-center/solana-too-ambitious-for-ethereum-targets-nasdaq-visa

19. Proof of History: How Solana brings time to crypto. (2021, November 30). *Solana News*. https://solana.com/news/proof-of-history

20. Tan, E. (2023, May 11). DISH to Tap Into Blockchain-Based Helium 5G Network. *CoinDesk*. https://www.coindesk.com/business/2021/10/26/dish-to-tap-into-blockchain-based-helium-5g-network

21. CoinMarketCap. (2022). What Is Helium (HNT)? Features, Tokenomics, and Price Prediction. *CoinMarketCap Alexandria*. https://coinmarketcap.com/alexandria/article/what-is-helium-hnt-features-tokenomics-and-price-prediction

22. Turck, M. (2022, March 25). *In Conversation with Amir Haleem*, CEO, Helium. Matt Turck. https://mattturck.com/helium2022/#more-1609

23. Roose, K. (2022, August 3). Helium Network Hints at Crypto's Practical Uses. *The New York Times*. https://www.nytimes.com/2022/02/06/technology/helium-cryptocurrency-uses.html

24. Helium. (2022, January 7). Literally, a Better Mousetrap from Helium & Victor® Powered by The People's Network. *Medium*. https://blog.helium.com/literally-a-better-mousetrap-from-helium-victor-powered-by-the-peoples-network-b3e7af96947b

25. Turck, M. (2022, March 25). *In Conversation with Amir Haleem*, CEO, Helium. Matt Turck. https://mattturck.com/helium2022/#more-1609

26. Ligon, C. (2023, May 11). Helium Becomes Nova Labs After Raising $200M in Fresh Capital. *CoinDesk*. https://www.coindesk.com/business/2022/03/30/helium-becomes-nova-labs-after-raising-200m-in-fresh-capital

27. Bambrough, B. (2022, May 12). $1 Trillion Crypto Meltdown—Huge Crash Wipes Out The Price Of Bitcoin, Ethereum, BNB, XRP, Cardano, Solana, Terra's Luna And Avalanche. *Forbes*. https://www.forbes.com/sites/billybambrough/2022/05/12/1-trillion-crypto-meltdown-huge-crash-wipes-out-the-price-of-bitcoin-ethereum-bnb-xrp-cardano-solana-terras-luna-and-avalanche/?sh=56caa6ed45fd

28. Howcroft, E. (2021, September 9). Set of "Bored Ape" NFTs sells for $24.4 mln in Sotheby's online auction. *Reuters*. https://www.reuters.com/lifestyle/set-bored-ape-nfts-sell-244-mln-sothebys-online-auction-2021-09-09

29. Crawley, J. (2023, May 11). NFT Trading Volume Surges 700% to $10.7B in Q3. *CoinDesk*. https://www.coindesk.com/business/2021/10/05/nft-trading-volume-surges-700-to-107b-in-q3

30. Top NFT Collections. (n.d.). DappRadar. Retrieved November 30, 2021, from https://dappradar.com/rankings/nft/collections

31. Rosenberg, L. (2022, February 5). Metaverse 101: Defining the key components. *VentureBeat*. https://venturebeat.com/business/metaverse-101-defining-the-key-components

32. Hamilton, I. A. (2022, April 12). Meta will charge creators fees of up to 47.5% to sell virtual wares in its metaverse. *Business Insider*. https://www.businessinsider.com/meta-metaverse-charges-creators-47-percent-sales-fee-cut-zuckerberg-2022-4

33. Robertson, H. (2021, December 27). A co-founder of The Sandbox explains what on earth is going on in the metaverse—where land is selling for millions and Snoop Dogg is hanging out. *Markets Insider*. https://markets.businessinsider .com/news/stocks/metaverse-the-sandbox-virtual-world-land-sales-decentraland-nfts-crypto-2021-12

About the Women

Stevie Sats is an author, editor and content creator. One of her favorite pastimes is sharing Bitcoin stories; the other is dancing. Her fintech training includes classes from The Crypto Academy and Copenhagen Business School, plus some coursework in SQL and cryptography. Prior to the world of blockchain, Stevie worked in education and e-commerce.

Maria Ortiz Spillane is a leader, tech marketer and seasoned brand strategist with a focus on Web 3.0, blockchain and crypto. She has an eye for the trends as well as what's making a lasting impact. Maria is passionate about sharing the potential of blockchain to transform the traditional ways of marketing, while helping women learn about this nascent space.

Maria and Stevie met in a virtual course through The Crypto Academy, where they connected with Bitcoiners and crypto enthusiasts from around the world. The strong network of women there empowered these two to reflect on their shared knowledge and start a newsletter.

www.ingramcontent.com/pod-product-compliance
Lightning Source LLC
Chambersburg PA
CBHW030509210326
41597CB00013B/844